Mathematical Theory
of Subdivision

Mathematical Theory of Subdivision

Finite Element and Wavelet Methods

Sandeep Kumar

Ashish Pathak

Debashis Khan

CRC Press

Taylor & Francis Group

Boca Raton London New York

CRC Press is an imprint of the
Taylor & Francis Group, an **informa** business

CRC Press
Taylor & Francis Group
6000 Broken Sound Parkway NW, Suite 300
Boca Raton, FL 33487-2742

© 2020 by Taylor & Francis Group, LLC
CRC Press is an imprint of Taylor & Francis Group, an Informa business

No claim to original U.S. Government works

Printed on acid-free paper

International Standard Book Number-13: 978-1-138-05158-4 (Hardback)

Library of Congress Cataloging-in-Publication Data
Names: Kumar, Sandeep (Professor of mechanical engineering), author. | Pathak, Ashish, author. | Khan, Debashis, author. Title: Mathematical theory of subdivision : finite element and wavelet methods / Sandeep Kumar, Ashish Pathak, Debashis Khan. Description: Boca Raton, Florida : CRC Press, [2019] | Includes bibliographical references and index. Identifiers: LCCN 2019008388 | ISBN 9781138051584 (hardback : alk. paper) | ISBN 9781315168265 (ebook) Subjects: LCSH: Finite element method. | Wavelets (Mathematics) | Numerical analysis. | Harmonic analysis. | Generalized spaces. Classification: LCC QC20.7.F56 K85 2019 | DDC 530.15/1825--dc23 LC record available at https://lccn.loc.gov/2019008388

Visit the Taylor & Francis Web site at
http://www.taylorandfrancis.com

and the CRC Press Web site at
http://www.crcpress.com

Printed and bound in Great Britain by
TJ International Ltd, Padstow, Cornwall

*Dedicated to all those who blessed us and
imparted their wisdom to make us wiser.*

Contents

Preface

SIMPLE LOGIC CANNOT EXPLAIN everything, and perfect generalization can be simplified to only a limited extent. The dilemma for God and the great mathematician is how to explain complexities in simple words. To keep it simple, God prefers to reveal partial truths acceptable to an individual. We scientists may sometimes reveal partial truths acceptable to a particular class (e.g., definition of vectors to engineers), but this partial revealing is sometimes to hide weakness as well. We are writing this book with the belief that simplicity on fundamental concepts is more useful than the hidden fundamentals within the complex advanced scientific theories. Students can easily understand advanced level books after understanding the fundamental concepts presented in this book.

We pay our respect to all great teachers who simplify subjects in wonderful ways. But sometimes we need to learn the subject ourselves due to various constraints. Subjects like the finite element method (FEM), wavelets and functional analysis are quite difficult for self-study, particularly if we are unlucky in getting a good fundamental book. While writing this book, we realized that nature is the best teacher. Let us see how nature teaches us. Suppose nature wants to explain to a child about mangos; it gives fruit to enjoy the taste. A child learns how to peel it in the next stage. Recognition of the tree and growing of trees are the parts of later stages. In the final stage, the child recognizes the advantages and disadvantages of eating mangoes. Unfortunately, most mathematics books do not like to give the taste of fruit easily. Their approach is to force one to do mind boggling exercises which often require the assistance of instructors. This book is presented in the three stages: (i) peeling the fruit to taste it, (ii) growing the tree to get the fruit and (iii) learning the conditions to know when it will not be fruitful.

In order to understand the third stage, which is a very necessary part of any knowledge, it is important to see the good and the bad, observe success and failure, simultaneously. These extreme ends help to avoid any kind of biased views on success and failure of any field. The success stories should be tagged with the cause of failure. The purpose is not to discourage but to raise questions on validity for perfect application of the method. Since this field contains many mathematical equations, it becomes too specialized and accessible only to a limited number of people who have a good mathematical background. We believe that some basic knowledge of functional analysis is essential to get a proper understanding of FEM and wavelets. Functional analysis, which is generally not taught in engineering courses, can only explain the causes of success and limitation of FEM and wavelets. For proper dependence on FEM and wavelets for various types of engineering problems, engineers should learn functional analysis. Functional analysis deals with the concepts of infinite dimensional spaces. FEM and wavelets are used to approximate infinite dimensional problems into solvable finite dimensional problems. Application of functional analysis concepts provides the logical continuity to reduce an infinite dimension problem to a solvable finite dimension problem.

The finite element method has achieved astonishing success in solving a certain class of partial differential equations in complex domain. Similarly, one can find every day a new success story of wavelets. Simplicity and versatility of these methods are so impressive that we get attracted as soon as we come across them and immediately we attempt to solve our problems. There are many good books which introduce the finite element method to students. There are many books on wavelets for students with different backgrounds, but they lack simple treatment on partial differential equations. The motivation of writing this book is to present these techniques in a simplest possible way by assuming no background of the reader in any field. Knowledge of linear algebra and partial differential equations, which is provided in undergraduate level in science and engineering, is sufficient to understand this book. The present book is to attract some intricate mathematical issues in simple language to people who do not have sufficient mathematical background. The study of the book will help practicing engineers and scientists in avoiding some fundamental mistakes which they often make due to lack of the knowledge. The book will not only raise some serious questions on the method but also introduce FEM and wavelets in a very simple way. To keep the book concise, we do not present many practical problems, but MATHEMATICA and MATLAB programs are frequently used, which can be easily modified to solve any type of problems.

This book will not only introduce these interesting areas for some application point of view but also present mathematical theories on them. The mathematical general theories available in abstract forms are not easily acceptable by everybody. Leave aside engineers and physicists; students of mathematics at the PG level do not enjoy subjects like functional analysis, measure theory or distribution theory. This book will aim to create interest on these mathematical theories among students by showing them the role of these theories in understanding the cause of success and failure of these great numerical methods. The challenge for the book is to present abstract mathematical concepts in simple and lucid form so too complex generalization is avoided.

The contents of the book are logically divided into seven chapters. Initially, FEM and wavelet methods are described with the help of some simple steps in the first two chapters. Both these chapters have solved numerical examples. The third and fourth chapters discuss operators and spaces which are required to understand the mathematical basis of FEM and wavelets. Mathematical theories of FEM and wavelets are explained in the subsequent two chapters with lucid explanation. Thereafter, the seventh chapter dealing with numerical error estimation techniques has been described with a few solved examples. We hope students will find this approach more appealing than the conventional system.

We would like to express our special thanks to the authors like Kevin Amaratunga, E. J. Stollnitz, Paul S. Addison, J. N. Reddy, Gilbert Strang, W. Sweldens and B. D. Reddy, among others, who inspired us by their research and simplicity in presentation. We want to thank all the former and present students without whose efforts this book could not have been written. We are also thankful to all the participants of our seminars and lectures who regularly provided the feedback on the presentation. We hope that the structure of presentation will provide the necessary fundamentals to students with a modest background in mathematics and generate lots of curiosity for future research and development.

We express our thanks to the staff at CRC press for assistance and expert guidance on preparation of the manuscript.

MATLAB® is a registered trademark of The MathWorks, Inc. For product information, please contact:

The MathWorks, Inc.
3 Apple Hill Drive
Natick, MA 01760-2098 USA
Tel: 508-647-7000
Fax: 508-647-7001
E-mail: info@mathworks.com
Web: www.mathworks.com

Authors

Sandeep Kumar serves as a professor in the Department of Mechanical Engineering at Indian Institute of Technology (Banaras Hindu University), Varanasi. He received his PhD from the Applied Mechanics Department, Indian Institute of Technology, Delhi, in 1999. His field of interest is computational mechanics: wavelets, finite element method, meshless method, etc.

Ashish Pathak serves as an assistant professor in the Department of Mathematics, Institute of Science (Banaras Hindu University). He received his PhD from the Department of Mathematics, Banaras Hindu University in 2009. His research interests include wavelet analysis, functional analysis and distribution theory.

Debashis Khan received his PhD in Mechanical Engineering from Indian Institute of Technology Kharagpur in 2007. Just after completing his PhD, he joined as an assistant professor in the Department of Mechanical Engineering at Indian Institute of Technology (Banaras Hindu University) Varanasi, and presently he is serving as associate professor in the same department. His research interests include solid mechanics, fracture mechanics, continuum mechanics, finite deformation plasticity and finite element method.

CHAPTER 1

Overview of Finite Element Method

THE FINITE ELEMENT METHOD IS known as one of the most effective computational techniques. The students are naturally curious about it and wish to understand it quickly. We choose not to present the exhaustive theory in the beginning. The basic steps involved in the finite element analysis of a problem are given in this chapter which will be sufficient to solve many problems. The taste of the basic steps will whet their appetite for the theory of finite element analysis which is presented in Chapter 5.

1.1 SOME COMMON GOVERNING DIFFERENTIAL EQUATIONS

Many physical phenomena can be described by using differential equations. The finite element method is one of the most powerful numerical methods to solve them [1–5]. Consider some differential equations which are often solved by engineers and scientists.

PROBLEM 1.1

A bar, shown in Figure 1.1, is subjected to uniform axial load f. The axial displacement u in the domain Ω ($x=0$ to L) can be obtained by solving the differential equation

$$-\frac{d}{dx}\left(EA\frac{du}{dx}\right)=f \tag{1.1a}$$

The boundary conditions are given as:

$$u(0)=0 \text{ and } u(L)=0 \tag{1.1b}$$

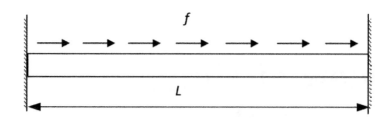

FIGURE 1.1   A bar with uniformly distributed load and fixed ends.

PROBLEM 1.2

Transverse deflection w in a beam, shown in Figure 1.2, is governed by the differential equation

$$\frac{d^2}{dx^2}\left(EI\frac{d^2w}{dx^2}\right)=f(x) \tag{1.2a}$$

Both the ends are fixed so the boundary conditions are:

$$\left(w\right)_{x=0}=\left(w\right)_{x=L}=\left(\frac{dw}{dx}\right)_{x=0}=\left(\frac{dw}{dx}\right)_{x=L}=0 \tag{1.2b}$$

PROBLEM 1.3

Deflection u of a membrane due to transverse load f is governed by Poisson's equation

$$-\nabla.\left(a\nabla u\right)=f \tag{1.3}$$

The gradient operator can be expressed as

$$\nabla=\hat{i}\frac{\partial}{\partial x}+\hat{j}\frac{\partial}{\partial y}+\hat{k}\frac{\partial}{\partial z}$$

These are some common differential equations which we often solve using FEM and whose solutions are very reliable, but there are some partial differential equations where we have to be more careful while applying this method. It is an interesting area of research and it is the subject of Chapter 5.

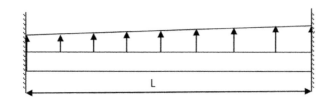

FIGURE 1.2   Beam with fixed ends and transverse load.

1.2 BASIC STEPS OF FINITE ELEMENT METHOD

The finite element procedures are used to solve simple as well as complex domain problems, which are governed by differential equations and boundary conditions. The procedure can be explained with the help of simple steps shown in Figure 1.3.

We use these FEM steps to solve every problem. For beginners, finding stiffness matrix may be a cumbersome process; therefore, initially we will use the relevant stiffness matrices and apply these steps to solve the problems. Every problem can be easily solved if the corresponding element stiffness matrix is available. The detailed procedure of finding element stiffness matrix is presented in Sections 1.3 through 1.5.

Step 1: Discretization of domain:

Generally, engineering problems have complex domain. Figures 1.1, 1.2, and 1.4 show the domain of the problems which are bar, beam and some arbitrary 2D domain. Every domain Ω of the continuum is discretized by imaginary lines or surfaces or volumes. Such division of region forms a set of "finite elements" which are simple and similar. For example, a 2D domain (3D domain with constant thickness) can be represented by a set of simple triangular elements as shown in Figure 1.5. The purpose of such division is to represent the field variable (displacement, temperature, etc.) of the differential equation of the whole domain with the help of finite number of constants. Each element consists of some nodes, primarily at its boundaries, which interconnect the other adjacent element—in this example, the three vertices form the three nodes of the element. The field variables of each element are uniquely defined using interpolation functions in terms of nodal variables. After discretization, the nodes and the elements are to be numbered in a proper order. To distinguish the two numberings, the element numbers are expressed within brackets, e.g., (1), (2), etc, and the node numbers are shown without brackets, e.g., 1, 2, etc. The accuracy of the solution increases as the number of elements within the domain or the number of nodes within the element is increased.

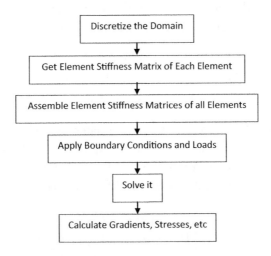

FIGURE 1.3   Overview of the finite element method.

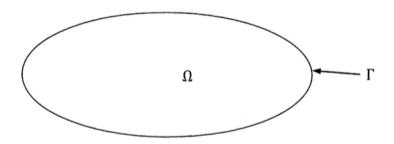

FIGURE 1.4   Domain Ω with boundary Γ for the Poisson's equation.

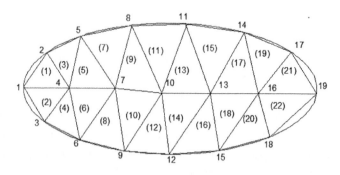

FIGURE 1.5   Discretization of 2D domain.

Step 2: Get element stiffness matrix:

Each differential equation with a particular type of element will provide a unique set of algebraic equations which can be expressed in terms of matrix known as element stiffness matrix $[k^e]$, nodal variables $\{u^e\}$ and nodal forces $\{f^e\}$ in the element.

$$[k^e]\{u^e\}=\{f^e\} \tag{1.4}$$

where superscript e denotes the element number. The proof of the matrix is presented in the Sections 1.3 through 1.5. A typical stiffness matrix of three noded triangular elements with single degree of freedom for each node can be expressed as:

$$\begin{bmatrix} k_{I,I}^e & k_{I,II}^e & k_{I,III}^e \\ k_{II,I}^e & k_{II,II}^e & k_{II,III}^e \\ k_{III,I}^e & k_{III,II}^e & k_{III,III}^e \end{bmatrix}\begin{Bmatrix} u_I^e \\ u_{II}^e \\ u_{III}^e \end{Bmatrix}=\begin{Bmatrix} f_I^e \\ f_{II}^e \\ f_{III}^e \end{Bmatrix} \tag{1.5}$$

The coefficients will depend on the coordinates of nodes as expressed by equation (1.40). For the element ($e = 5$) shown in Figure 1.5, the local numbers I, II, III are assigned to the global node numbers 5, 4, 7. Local numbering in anticlockwise direction is used for every element. Therefore stiffness matrix can be expressed as

$$[k^e]=[k_{i,j}^e]=\begin{bmatrix} k_{5,5}^5 & k_{5,4}^5 & k_{5,7}^5 \\ k_{4,5}^5 & k_{4,4}^5 & k_{4,7}^5 \\ k_{7,5}^5 & k_{7,4}^5 & k_{7,7}^5 \end{bmatrix} \tag{1.6}$$

Similarly, local nodal variables and forces can be transformed to global nodal variables and forces. Initially, we are learning how to use the element stiffness matrices. The finite element software also selects the appropriate stiffness matrix from its library and solves the problem. The element stiffness matrix for a bar and beam are given by equations (1.11) and (1.54), respectively.

Step 3: Assembly of stiffness matrices:

For n nodes in the domain of single degree of freedom system, the size of global stiffness matrix will be $n \times n$. Each element $k_{i,j}^e$ of the element stiffness matrix $[k^e]$ will be added at their i^{th} row and j^{th} column of the global stiffness matrix. In this case,

$$[K] = \begin{bmatrix} 0 & 0 & 0 & 0 & 0 & 0 & 0 & . & . & 0 \\ 0 & 0 & 0 & 0 & 0 & 0 & 0 & . & . & 0 \\ 0 & 0 & 0 & 0 & 0 & 0 & 0 & . & . & 0 \\ 0 & 0 & 0 & k_{4,4}^5 & k_{4,5}^5 & 0 & k_{4,7}^5 & . & . & 0 \\ 0 & 0 & 0 & k_{5,4}^5 & k_{5,5}^5 & 0 & k_{5,7}^5 & . & . & 0 \\ 0 & 0 & 0 & 0 & 0 & 0 & 0 & . & . & 0 \\ 0 & 0 & 0 & k_{7,4}^5 & k_{7,5}^5 & 0 & k_{7,7}^5 & . & . & 0 \\ . & . & . & . & . & . & . & . & . & 0 \\ . & . & . & . & . & . & . & . & . & 0 \\ 0 & 0 & 0 & 0 & 0 & 0 & 0 & . & . & 0 \end{bmatrix} \tag{1.7}$$

Similarly, other element stiffness matrices will be added in the global matrix at their respective places. The forces at each node due to each element are also to be added. The assembled equation can be expressed as

$$[K]\{u\} = \{F\} \tag{1.8}$$

Step 4: Solution:

Any standard method can be used to solve it after substituting prescribed essential boundary conditions in terms of nodal variables. For example, for nodal variable $u_i = 0$, the ith row and column are deleted from the set of equation and the reduced set of equations are to be solved. For $u_i = \alpha$ (constant), add a very large number $C = \max |k_{ij}| \times 10^4$ at global stiffness matrix coefficient k_{ii} and substitute $C \times \alpha$ at i^{th} node of the load vector. Due to substitution of a very large value, all the terms of the ith row become negligible and the equation of this row reduces to $C \times u_i = C \times \alpha$. The matrix is generally very large but sparse (large number of coefficients matrix are zero). The proper node numbering during discretization generates a banded form of matrix. Such a banded matrix is stored and solved efficiently by a suitable algorithm. Instead of direct solvers, iterative matrix solvers are used for a very large matrix. Often a preconditioner is used to improve the convergence of iterative matrix solvers.

Step 5: Calculation of gradients and stresses, etc.:

The analysis of results, called post processing, is often necessary. For example, to study of the failure of structures, it is necessary to examine strain and stress which can be obtained by using derivative of the known displacement field and constitutive equations, respectively. Sometimes, deformed shape is displayed over the undeformed shape of the structures. Such post processing is very well utilized in the commercial FEM software.

Solution of Problem 1.1:

Equation (1.1) is the governing equation of many problems such as bars, cables, heat transfer, flow in pipe, etc. In this exercise, we consider it for axial deformation of the bar shown in Figure 1.6. The steel bar is fixed at both the ends and boundary conditions are: $u|_{x=0}=0$ and $u|_{x=1}=0$. A uniform traction force $f=1$ N/m is acting on it. We want to find the reactions at the ends. Young's modulus of the material is $E=2.0$ GPa. The bar is discretized using four elements of equal length as shown in Figure 1.6.

For element (2), the global node numbers (2 and 3) and the local node numbers (I and II) are shown in Figure 1.6c.

Since f is uniformly distributed load per unit length, the total load on the element of length l_2 is fl_2. The procedure of distributing this load on nodes is discussed in Section 1.3, but the load will be equally distributed in case of two nodes. The stiffness matrix and force vectors are obtained using the local coordinates in Section 1.3 as

$$\begin{bmatrix} k_{I,I}^e & k_{I,II}^e \\ k_{II,I}^e & k_{II,II}^e \end{bmatrix} \begin{Bmatrix} u_I^e \\ u_{II}^e \end{Bmatrix} = \begin{Bmatrix} f_I^e \\ f_{II}^e \end{Bmatrix}$$

where $k_{I,I}^e = k_{II,II}^e = k_e$; $k_{II,I}^e = k_{I,II}^e = -k_e$; $k_e = \frac{E_e A_e}{l_e} = 8 \times 10^5$ and $f_I^e = f_{II}^e = \frac{fl_e}{2} = 0.125$.

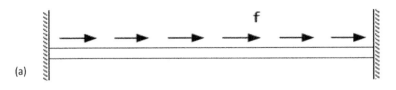

FIGURE 1.6   A bar and its discretization: (a) A bar of uniform cross-sectional area A = 1 cm² and length L = 1 m with fixed ends. (b) Discretization of bar. (c) Element (2) of bar.

It is not difficult to understand the derivation of element stiffness matrix of a bar; therefore, students can read Section 1.3 before proceeding for assembly. This equation for element 1 is expressed as:

$$k_1 \begin{bmatrix} 1 & -1 \\ -1 & 1 \end{bmatrix} \begin{Bmatrix} u_1 \\ u_2 \end{Bmatrix} = \begin{Bmatrix} f_1^1 \\ f_2^1 \end{Bmatrix}$$

Similarly, the equation for element 2 is expressed as:

$$k_2 \begin{bmatrix} 1 & -1 \\ -1 & 1 \end{bmatrix} \begin{Bmatrix} u_2 \\ u_3 \end{Bmatrix} = \begin{Bmatrix} f_2^2 \\ f_3^2 \end{Bmatrix}$$

The equation for element 3 is expressed as:

$$k_3 \begin{bmatrix} 1 & -1 \\ -1 & 1 \end{bmatrix} \begin{Bmatrix} u_3 \\ u_4 \end{Bmatrix} = \begin{Bmatrix} f_3^3 \\ f_4^3 \end{Bmatrix}$$

The equation for element 4 is expressed as:

$$k_4 \begin{bmatrix} 1 & -1 \\ -1 & 1 \end{bmatrix} \begin{Bmatrix} u_4 \\ u_5 \end{Bmatrix} = \begin{Bmatrix} f_4^4 \\ f_5^4 \end{Bmatrix}$$

Sometimes, the nodes are subjected to some external concentrated load. In this example, there is no concentrated force at nodes 2, 3, and 4, but due to reactions at the supports 1 and 5, the nodes will get external forces R_1 and R_2. These unknown forces are added in the assembled equation. There are 5 nodes, hence the size of global stiffness matrix is 5 × 5. The element stiffness matrices are assembled in the global stiffness matrix. The final equation can be expressed as:

$$\begin{bmatrix} k_1 & -k_1 & 0 & 0 & 0 \\ -k_1 & k_1+k_2 & -k_2 & 0 & 0 \\ 0 & -k_2 & k_2+k_3 & -k_3 & 0 \\ 0 & 0 & -k_3 & k_3+k_4 & -k_4 \\ 0 & 0 & 0 & -k_4 & k_4 \end{bmatrix} \begin{Bmatrix} u_1 \\ u_2 \\ u_3 \\ u_4 \\ u_5 \end{Bmatrix} = \begin{Bmatrix} f_1^1 \\ f_2^1+f_2^2 \\ f_3^2+f_3^3 \\ f_4^3+f_4^4 \\ f_5^4 \end{Bmatrix} + \begin{Bmatrix} R_1 \\ 0 \\ 0 \\ 0 \\ R_2 \end{Bmatrix}$$

Since $u_1 = 0$ and $u_5 = 0$, the first and fifth rows and columns are eliminated. Substituting all the known values we get

$$\begin{bmatrix} 16 & -8 & 0 \\ -8 & 16 & -8 \\ 0 & -8 & 16 \end{bmatrix} \begin{bmatrix} u_2 \\ u_3 \\ u_4 \end{bmatrix} = \begin{bmatrix} 0.25 \\ 0.25 \\ 0.25 \end{bmatrix} + \begin{bmatrix} 0 \\ 0 \\ 0 \end{bmatrix}$$

We find the displacements from the matrix as $u_2 = 4.68 \times 10^{-7}$ m, $u_3 = 6.25 \times 10^{-7}$ m, $u_4 = 4.68 \times 10^{-7}$ m.

Reactions are computed from the eliminated equations

$$R_1 = k_1 u_1 - k_1 u_2 - f_1^1$$

$$R_2 = k_4 u_5 - k_4 u_4 - f_5^4$$

The solutions are $R_1 = -0.5$ N and $R_2 = -0.5$ N.

Solution of Problem 1.2:

Similar to the previous example, we can easily solve the problem of a beam (Figure 1.7) by using the element stiffness matrix (derived in Section 1.5). The beam is subjected to uniformly distributed load $f_0 = 5$ kN/m and a concentrated load $F_0 = 10$ kN at a distance of $l/4$ from the left end.

For simplicity, the beam is represented by two elements with a node created at $x = l/4$ for the concentrated load. The beam element is shown in the Figure 1.7c. The field variable of interest is the transverse deflection w and rotation θ at the nodes. The displacements and rotations are measured from its straight undeflected position.

The algebraic equation for an element (equations 1.54 and 1.56) is

$$\frac{2E_e I_e}{l_e^3} \begin{bmatrix} 6 & 3l_e & -6l_e & 3l_e \\ 3l_e & 2l_e^2 & -3l_e & l_e^2 \\ -6 & -3l_e & 6 & -3l_e \\ 3l_e & l_e^2 & -3l_e & 2l_e^2 \end{bmatrix} \begin{bmatrix} w_I \\ \theta_I \\ w_{II} \\ \theta_{II} \end{bmatrix} = \frac{f l_e}{12} \begin{bmatrix} 6 \\ l_e \\ 6 \\ -l_e \end{bmatrix}$$

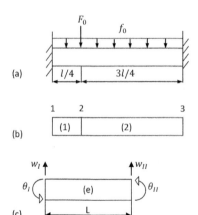

w_I, w_{II} are transverse deflections at nodes.

θ_I, θ_{II} are slopes at nodes.

FIGURE 1.7   Beam and its discretization: (a) Beam with transverse force. (b) Discretization of beam. (c) Beam element.

In this case of two elements of length $l/4$ and $3l/4$, the element stiffness matrices are

$$\left[K^1\right] = \frac{128EI}{l^3}\begin{bmatrix} 6 & \dfrac{3l}{4} & -\dfrac{6l}{4} & \dfrac{3l}{4} \\[2mm] \dfrac{3l}{4} & \dfrac{2l^2}{16} & -\dfrac{3l}{4} & \dfrac{l^2}{16} \\[2mm] -6 & -\dfrac{3l}{4} & 6 & -\dfrac{3l}{4} \\[2mm] \dfrac{3l}{4} & \dfrac{l^2}{16} & -\dfrac{3l}{4} & \dfrac{2l^2}{16} \end{bmatrix} ; \left[K^2\right] = \frac{128EI}{27l^3}\begin{bmatrix} 6 & \dfrac{9l}{4} & -\dfrac{18l}{4} & \dfrac{9l}{4} \\[2mm] \dfrac{9l}{4} & \dfrac{18l^2}{16} & -\dfrac{9l}{4} & \dfrac{9l^2}{16} \\[2mm] -6 & -\dfrac{9l}{4} & 6 & -\dfrac{9l}{4} \\[2mm] \dfrac{9l}{4} & \dfrac{9l^2}{16} & -\dfrac{9l}{4} & \dfrac{18l^2}{16} \end{bmatrix}$$

and load vectors are

$$\{F^1\} = \frac{fl}{48}\begin{Bmatrix} 6 \\ l \\ 6 \\ -l \end{Bmatrix} \{F^2\} = \frac{fl}{16}\begin{Bmatrix} 6 \\ l \\ 6 \\ -l \end{Bmatrix},$$

The final equation is obtained after assembly,

$$\frac{128EI}{l^3}\begin{bmatrix} 6 & \dfrac{3l}{4} & -\dfrac{6l}{4} & \dfrac{3l}{4} & 0 & 0 \\[2mm] \dfrac{3l}{4} & \dfrac{2l^2}{16} & -\dfrac{3l}{4} & \dfrac{l^2}{16} & 0 & 0 \\[2mm] -6 & -\dfrac{3l}{4} & \left(6 + \dfrac{6}{27}\right) & \left(-\dfrac{3l}{4} + \dfrac{9l}{108}\right) & -\dfrac{18l}{108} & \dfrac{9l}{108} \\[2mm] \dfrac{3l}{4} & \dfrac{l^2}{16} & -\left(\dfrac{3l}{4} + \dfrac{9l}{108}\right) & \left(\dfrac{2l^2}{16} + \dfrac{18l^2}{432}\right) & -\dfrac{9l}{108} & \dfrac{9l^2}{432} \\[2mm] 0 & 0 & -\dfrac{6}{27} & -\dfrac{9l}{108} & \dfrac{6}{27} & -\dfrac{9l}{108} \\[2mm] 0 & 0 & \dfrac{9l}{108} & \dfrac{9l^2}{432} & \dfrac{9l}{108} & \dfrac{18l^2}{432} \end{bmatrix} \begin{Bmatrix} w_1 \\ \theta_1 \\ w_2 \\ \theta_2 \\ w_3 \\ \theta_3 \end{Bmatrix} = \frac{fl}{16}\begin{Bmatrix} 6/3 \\ l/3 \\ 8 \\ 2l/3 \\ 6 \\ -l \end{Bmatrix} + \begin{Bmatrix} F_1 \\ M_1 \\ 10 \\ 0 \\ F_3 \\ M_3 \end{Bmatrix}$$

Due to fixed support, F_1, M_1, F_3, and M_3 are unknown reaction forces and moments. The constraint conditions are $w_1 = \theta_1 = w_3 = \theta_3 = 0$. Therefore, after elimination of corresponding 1st, 2nd, 5th, and 6th rows and columns, we get the algebraic equations in the matrix form as

$$\begin{bmatrix} \left(6 + \dfrac{6}{27}\right) & \left(-\dfrac{3l}{4} + \dfrac{9l}{108}\right) \\[3mm] -\left(\dfrac{3l}{4} + \dfrac{9l}{108}\right) & \left(\dfrac{2l^2}{16} + \dfrac{18l^2}{432}\right) \end{bmatrix} \begin{Bmatrix} w_2 \\ \theta_2 \end{Bmatrix} = \begin{Bmatrix} -\dfrac{fl}{2} \\[3mm] -\dfrac{fl}{24} \end{Bmatrix} + \begin{Bmatrix} 10 \\ 0 \end{Bmatrix}$$

For $l = 1$ m, the deflection and slope at node 2 are obtained as

$$w_2 = 2.516; \qquad \theta_2 = 10.306$$

The reaction forces and moments can be obtained by substitution of w_2 and θ_2 in the assembly.

Solution of Problem 1.3:

Poisson's equation is used in many science and engineering problems such as in heat transfer, torsion, irrotational flow, electro and magnetostatics, etc. We want to find transverse deflection u of a membrane within a rectangular domain due to uniformly distributed transverse load f_0. The deformation is governed by Poisson's equation:

$$-\nabla^2 u = f_0 \;\; \text{or} \; -\left(\frac{\partial^2 u}{\partial x^2} + \frac{\partial^2 u}{\partial y^2} \right) = f_0$$

In Ω

$$\Omega = \left\{ (x,y): -l < x < l, -l < y < l \right\}$$

The four edges of the membrane are clamped such that $u = 0$ on Γ (Figure 1.8).
   On the line of symmetry, the normal derivative of the solution

$$q_n = \frac{\partial u}{\partial n} = \frac{\partial u}{\partial x} n_x + \frac{\partial u}{\partial y} n_y = 0$$

owing to symmetry along the diagonal we model the triangular domain as shown in Figure 1.8c. We use a uniform mesh of four linear triangular elements to represent the domain. Now consider triangular element (1) as shown in the figure. The element coefficient matrix and force vector are (see equations 1.40 and 1.42)

$$[K^1] = \frac{1}{2ab} \begin{bmatrix} b^2 & -b^2 & 0 \\ -b^2 & a^2 + b^2 & -a^2 \\ 0 & -a^2 & a^2 \end{bmatrix}, \;\; \{f^1\} = \frac{f_0 ab}{6} \begin{Bmatrix} 1 \\ 1 \\ 1 \end{Bmatrix}$$

where a and b are the sides of a right angle triangle in which the right angle is at node II (local node number) and the diagonal line of the triangle connects node III to node I (anticlockwise direction). The same pattern for local node numbering is followed in all the elements.
   In this case $a = b = l/2 = 0.5$

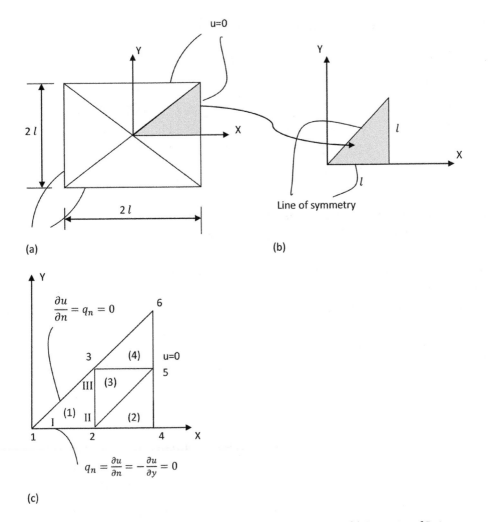

(c)

FIGURE 1.8 Domain of Poisson equation and its discretization: (a) Domain of Poisson equation. (b) Only 1/8 part of the domain due to symmetry. (c) Discretization of domain.

We can see in Figure 1.8c that, the elements (1), (2) and (4) are identical in orientation as well as geometry. Element (3) is geometrically identical to element (1) except that it is oriented differently. If we number the local nodes of element (3) to match those of element (1) then all four elements have the same element matrices, i.e.,

$$[K^1] = [K^2] = [K^3] = [K^4], \text{ and the force vector } \{f^1\} = \{f^2\} = \{f^3\} = \{f^4\}$$

With

$$[K^e] = \frac{1}{2}\begin{bmatrix} 1 & -1 & 0 \\ -1 & 2 & -1 \\ 0 & -1 & 1 \end{bmatrix}, \quad \{f^e\} = \frac{f_0}{24}\begin{Bmatrix} 1 \\ 1 \\ 1 \end{Bmatrix}$$

Due to six nodes, the size of the assembled coefficient matrix is 6×6.

$$\frac{1}{2}\begin{bmatrix} 1 & -1 & 0 & 0 & 0 & 0 \\ -1 & 4 & -2 & -1 & 0 & 0 \\ 0 & -2 & 4 & 0 & -2 & 0 \\ 0 & -1 & 0 & 2 & -1 & 0 \\ 0 & 0 & -2 & -1 & 4 & -1 \\ 0 & 0 & 0 & 0 & -1 & 1 \end{bmatrix}\begin{Bmatrix} U_1 \\ U_2 \\ U_3 \\ U_4 \\ U_5 \\ U_6 \end{Bmatrix} = \frac{f_0}{24}\begin{Bmatrix} 1 \\ 3 \\ 3 \\ 1 \\ 3 \\ 1 \end{Bmatrix} + \begin{Bmatrix} Q_1^1 \\ Q_2^1 + Q_3^2 + Q_1^3 \\ Q_3^1 + Q_2^2 + Q_1^4 \\ Q_2^3 \\ Q_1^2 + Q_3^3 + Q_2^4 \\ Q_3^4 \end{Bmatrix}$$

We assume that the nodal displacements, $U_4 = U_5 = U_6 = 0$.

And due to symmetry [4], the flux $Q_1^1 = Q_2^1 + Q_3^2 + Q_1^3 = Q_3^1 + Q_2^2 + Q_1^4 = 0$

Also assume $f_0 = 1$.

The preceding matrix can be solved for unknown quantities by using Penalty approach method.

Adding a large number $C = |\max k_{ij}| \times 10^4$ to the k_{44}, k_{55}, k_{66} and force vector on the right side.

Now $C = |\max k_{ij}| \times 10^4 = 4 \times 10^4$

After adding C and substituting values of other variables, the above matrix becomes

$$\frac{1}{2}\begin{bmatrix} 1 & -1 & 0 & 0 & 0 & 0 \\ -1 & 4 & -2 & -1 & 0 & 0 \\ 0 & -2 & 4 & 0 & -2 & 0 \\ 0 & -1 & 0 & 2+4\times10^4 & -1 & 0 \\ 0 & 0 & -2 & -1 & 4+4\times10^4 & -1 \\ 0 & 0 & 0 & 0 & -1 & 1+4\times10^4 \end{bmatrix}\begin{Bmatrix} U_1 \\ U_2 \\ U_3 \\ U_4 \\ U_5 \\ U_6 \end{Bmatrix} = \frac{1}{24}\begin{Bmatrix} 1 \\ 3 \\ 3 \\ 1 \\ 3 \\ 1 \end{Bmatrix} + \begin{Bmatrix} 0 \\ 0 \\ 0 \\ Q_2^3 + 4\times10^4 \\ Q_1^2 + Q_3^3 + Q_2^4 + 4\times10^4 \\ Q_3^4 + 4\times10^4 \end{Bmatrix}$$

Now

$$Q_2^3 + 4\times10^4 \approx 4\times10^4, \; Q_1^2 + Q_3^3 + Q_2^4 + 4\times10^4 \approx 4\times10^4, \; Q_3^4 + 4\times10^4 \approx 4\times10^4$$

$$2+4\times10^4 \approx 4\times10^4, \; 4+4\times10^4 = 4\times10^4, \; 1+4\times10^4 \approx 4\times10^4$$

Then

$$\frac{1}{2}\begin{bmatrix} 1 & -1 & 0 & 0 & 0 & 0 \\ -1 & 4 & -2 & -1 & 0 & 0 \\ 0 & -2 & 4 & 0 & -2 & 0 \\ 0 & -1 & 0 & 4\times10^4 & -1 & 0 \\ 0 & 0 & -2 & -1 & 4\times10^4 & -1 \\ 0 & 0 & 0 & 0 & -1 & 4\times10^4 \end{bmatrix}\begin{Bmatrix} U_1 \\ U_2 \\ U_3 \\ U_4 \\ U_5 \\ U_6 \end{Bmatrix} = \frac{1}{24}\begin{Bmatrix} 1 \\ 3 \\ 3 \\ 1 \\ 3 \\ 1 \end{Bmatrix} + \begin{Bmatrix} 0 \\ 0 \\ 0 \\ 4\times10^4 \\ 4\times10^4 \\ 4\times10^4 \end{Bmatrix}$$

The preceding matrix can now be easily solved to find U_1, U_2, U_3.

$$U_1 = 0.3120, U_2 = 0.2290, U_3 = 0.1770$$

1.3 ELEMENT STIFFNESS MATRIX FOR A BAR

1.3.1 Direct Method

A bar is a one-dimensional element used in truss or frame structure and supports an axial load only. For simplicity, we assume deformation as directly proportional to the axial load. The constant of proportionality is called as stiffness k. For uniform cross-section bar

$$k = \frac{EA}{l} \tag{1.9}$$

where E is the modulus of elasticity, A is the area of cross-section, and l is the length. Figure 1.9 shows an element of a bar.

At the equilibrium $f_I = -f_{II}$ and it can be written as

$$f_I = k(u_I - u_{II}) \tag{1.10a}$$

$$f_{II} = k(u_{II} - u_I) \tag{1.10b}$$

It can be expressed in the matrix form as

$$\begin{bmatrix} k & -k \\ -k & k \end{bmatrix} \begin{bmatrix} u_I \\ u_{II} \end{bmatrix} = \begin{bmatrix} f_I \\ f_{II} \end{bmatrix} \tag{1.11a}$$

or

$$\begin{bmatrix} k^e \end{bmatrix} \begin{bmatrix} u^e \end{bmatrix} = \begin{bmatrix} f^e \end{bmatrix} \tag{1.11b}$$

This element stiffness matrix is used in Example 1.1.

1.3.2 Potential Energy Method

Another method, which is generally preferred by solid mechanics books, is the principle of minimum potential energy (I). Potential energy of the bar under axial deformation can be expressed as

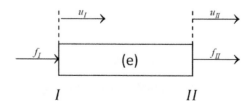

FIGURE 1.9   Bar with nodal forces and nodal displacements.

$$I = \text{Strain energy} - \text{work done by external forces}$$

$$= \frac{1}{2}k(u_{II} - u_I)^2 - (f_I u_I + f_{II} u_{II}) \tag{1.12}$$

The bar is in equilibrium, therefore the principle of minimum potential energy

$$\frac{\partial I}{\partial u_I} = \frac{\partial I}{\partial u_{II}} = 0$$

yields

$$k(u_I - u_{II}) = f_I \tag{1.13a}$$

$$k(u_{II} - u_I) = f_{II} \tag{1.13b}$$

This is identical to equation (1.11).

1.3.3 Higher Order Bar Element via Interpolation Functions

We cannot use these direct and simple methods in complex problems to find the element stiffness matrix. A generalized procedure uses functional (or in this case potential energy) from a differential equation as discussed in Section 5.3. For the differential equation (1.1), the expression for the potential energy is obtained as:

$$I = \frac{1}{2}\int_L EA\left(\frac{du}{dx}\right)^2 dx - \int_L uf dx - \sum_i u_i F_i \tag{1.14}$$

The first term on the right side is due to strain energy, and the remaining terms are work done by the forces. Here f is the distributed load, u_i is the displacement and F_i is the concentrated load at $x = x_i$. After finite element discretization, the potential energy can be expressed as

$$I = \frac{1}{2}\sum_e \int_{l_e} E^e A^e \left(\frac{du^e}{dx}\right)^2 dx - \sum_e \int_{l_e} u^e f^e dx - \sum_e u_i F_i \tag{1.15}$$

The first term generates element stiffness matrix and the remaining terms generate force vector.

The continuous field variable (in this case displacement field) $u^e(x)$ for an element can be assumed as linear function

$$u^e(x) = a_1 + a_2 x \tag{1.16}$$

or a higher order polynomial. For simplicity, we assume a quadratic polynomial

$$u^e(x) = a_1 + a_2 x + a_3 x^2 \tag{1.17}$$

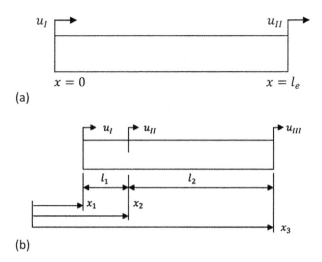

FIGURE 1.10 Bar elements: (a) Two-node bar element and (b) three-node bar element.

In the finite element method, it is essential to replace the unknown variables a_i of interpolation function equations (1.16) and (1.17) by unknown nodal displacement u_i (Figure 1.10).

Let $u = u_I$ at $x = 0$ and $u = u_{II}$ at $x = l_e$ then $a_1 = u_I$ and $a_2 = (u_{II} - u_I)/l_e$

Hence, the interpolation function can be expressed as

$$u^e(x) = \left(1 - \frac{x}{l_e}\right)u_I + \frac{x}{l_e}u_{II} \tag{1.18a}$$

or

$$u^e(x) = N_I u_I + N_{II} u_{II} \tag{1.18b}$$

Similarly, the interpolation function equation (1.17) can be expressed for an element with three nodes.

The first and last nodes are placed at the ends of element, and the other nodes can be assumed anywhere inside the element.

The interpolation function equation (1.17) can be expressed in terms of nodal displacement by satisfying

$$u^e(\text{at } x = x_1) = u_I; \; u^e(\text{at } x = x_2) = u_{II} \text{ and } u^e(\text{at } x = x_3) = u_{III}$$

For simplicity, we assume $x_1 = 0$, $x_2 = \frac{l}{2}$ and $x_3 = l$ then the coefficients a_i can be obtained from the following set of equations

$$u_I = a_1 + a_2.0 + a_3.0 \tag{1.19a}$$

$$u_{II} = a_1 + a_2\,\frac{l}{2} + a_3\left(\frac{l}{2}\right)^2 \tag{1.19b}$$

$$u_{III} = a_1 + a_2 l + a_3 (l)^2 \qquad (1.19c)$$

These three equations transform the interpolation function in equation (1.17) as

$$u^e(x) = \frac{2}{l^2}\left\{(x-l/2)(x-l)u_I + 2x(l-x)u_{II} + x(x-l/2)u_{III}\right\} \qquad (1.20a)$$

$$= N_I u_I + N_{II} u_{II} + N_{III} u_{III} \qquad (1.20b)$$

$$= [N]\{u^e\} \qquad (1.20c)$$

where $\{u^e\} = \{u_I, u_{II}, u_{III}\}^T$. Equations (1.18) and (1.20) are known as Lagrange interpolation functions. N_I, N_{II}, etc., are called as shape functions which have the following properties

$$N_i = \begin{cases} 1 & \text{at node } i \\ 0 & \text{at all other nodes} \end{cases} \qquad (1.21a)$$

and

$$\sum_e N_i = 1 \qquad (1.21b)$$

This property is valid for two and three-dimensional Lagrange interpolation functions. For quadratic element, we will substitute the interpolation function in the expression of potential energy.

$$\frac{du^e}{dx} = \frac{d}{dx}[N]\{u^e\} \qquad (1.22a)$$

$$= \left[\frac{dN_I}{dx}, \ \frac{dN_{II}}{dx}, \ \frac{dN_{III}}{dx}\right]\{u^e\} \qquad (1.22b)$$

$$= \frac{2}{l_e^2}\left[\frac{(4x-3l_e)}{2}, \ (2l_e - 4x), \ \frac{(4x-l_e)}{2}\right]\{u^e\} \qquad (1.22c)$$

$$= [B]\{u^e\} \qquad (1.22d)$$

Strain energy within an element $I = \dfrac{1}{2}\displaystyle\int_{l_e} E^e A^e \left(\dfrac{du^e}{dx}\right)^2 dx$

$$= \frac{E^e A^e}{2}\int \{u^e\}^T [B]^T [B]\{u^e\} dx \qquad (1.23a)$$

$$= \left\{u^e\right\}^{\mathrm{T}} \left(\frac{E^e A^e}{2} \int_{x_I}^{x_{II}} [B]^T [B] dx \right) \left\{u^e\right\} \tag{1.23b}$$

$$= \left\{u^e\right\}^{\mathrm{T}} \frac{E^e A^e}{2} \int_{x_I}^{x_{II}} \begin{bmatrix} \dfrac{dN_I}{dx} \dfrac{dN_I}{dx} & \dfrac{dN_I}{dx} \dfrac{dN_{II}}{dx} & \dfrac{dN_I}{dx} \dfrac{dN_{III}}{dx} \\[2ex] \dfrac{dN_{II}}{dx} \dfrac{dN_I}{dx} & \dfrac{dN_{II}}{dx} \dfrac{dN_{II}}{dx} & \dfrac{dN_{II}}{dx} \dfrac{dN_{III}}{dx} \\[2ex] \dfrac{dN_{III}}{dx} \dfrac{dN_I}{dx} & \dfrac{dN_{III}}{dx} \dfrac{dN_{II}}{dx} & \dfrac{dN_{III}}{dx} \dfrac{dN_{III}}{dx} \end{bmatrix} dx \left\{u^e\right\} \tag{1.23c}$$

$$= \frac{1}{2} \left\{u^e\right\}^{\mathrm{T}} \left[k^e\right] \left\{u^e\right\} \tag{1.23d}$$

In this case, element stiffness is

$$\left[k^e\right] = \frac{E^e A^e}{3l_e} \begin{bmatrix} 7 & -8 & 1 \\ -8 & 16 & -8 \\ 1 & -8 & 7 \end{bmatrix} \tag{1.24}$$

Similarly, the work done within an element by the distributed load can be expressed as

$$\int ufdx = \int_{x_I}^{x_{II}} \left\{u^e\right\}^T [N]^T fdx \tag{1.25a}$$

$$= \left\{u^e\right\}^T \begin{bmatrix} \displaystyle\int_{x_I}^{x_{II}} N_I fdx \\[2ex] \displaystyle\int_{x_I}^{x_{II}} N_{II} fdx \\[2ex] \displaystyle\int_{x_I}^{x_{II}} N_{III} fdx \end{bmatrix} = \left\{u^e\right\}^T \frac{fl}{6} \begin{Bmatrix} 1 \\ 4 \\ 1 \end{Bmatrix} \tag{1.25b}$$

The total potential energy is expressed as

$$I = \sum_e \frac{1}{2} \left\{u^e\right\}^T \left[k^e\right] \left\{u^e\right\} - \sum_e \left\{u^e\right\}^T \left\{F^e\right\} - \sum_e u_i F_i \tag{1.26a}$$

The assembled form of matrix can be expressed as

$$I = \frac{1}{2} \left\{u\right\}^T [k]\left\{u\right\} - \left\{u\right\}^T [F] \tag{1.26b}$$

where $\{u\}$ is the global displacement vector, $[k]$ is the global stiffness matrix and $\{F\}$ is the global load vector due to distributed loads. u_i and F_i are displacement and concentrated load at node i, respectively. The minimization of potential energy

$$\frac{\partial I}{\partial u_i} = 0 \quad i = 1, 2, \ldots N$$

leads to the set of the algebraic equations which can be expressed in the matrix form as

$$[k]\{u\} = \{F\}$$

The student should get the stiffness matrix for linear element using equation (1.18) which will be the same as matrix equation (1.11).

1.4 ELEMENT STIFFNESS MATRIX FOR SINGLE VARIABLE 2D ELEMENT

For a membrane of thickness t, the Poisson equation can be expressed by following equivalent statement (see Section 5.3): Find $u(x)$ that minimizes the functional. (Functional is defined in Chapter 4.)

$$I = t \iint_{\Omega} \left\{ \frac{1}{2}\left[\left(\frac{\partial u}{\partial x}\right)^2 + \left(\frac{\partial u}{\partial y}\right)^2 \right] - fu \right\} d\Omega \tag{1.27a}$$

After discretization, it can be expressed as

$$= \sum_e t \iint_{\Omega_e} \left\{ \frac{1}{2}\left[\left(\frac{\partial u^e}{\partial x}\right)^2 + \left(\frac{\partial u^e}{\partial y}\right)^2 \right] - u^e \times f \right\} d\Omega \tag{1.27b}$$

The procedure of finding stiffness matrix of this single variable two-dimensional problem is similar to what we discussed in Section 1.3.3. The two-dimensional domain Ω with boundary Γ are subdivided by using various types of two-dimensional triangular and quadrilateral elements, shown in Figure 1.11.

The nodes are placed at the ends of the edges, inside the edge (in case of six and eight node elements) and inside the domain (in case of nine node element) which need not be at the centre of the edge or domain. There are mainly two questions for such elements: how to select interpolation functions and how to numerically differentiate and integrate them for element stiffness matrix. Generally Pascal triangle is used to ensure that independent variables (x-, y- and z-coordinates) have an equal weight in the polynomial. For example, quadrilateral element with four nodes will be

$$u^e(x, y) = a_0 + a_1 x + a_2 y + a_3 xy \tag{1.28}$$

The fourth term cannot be $a_3 x^2$ or $a_3 y^2$ because the polynomial will not be symmetric in such case. The constant term a_0 is essential for rigid body displacement, and linear terms $a_1 x$ and $a_2 y$ are necessary for constant strain deformation. Mapping from x–y space to

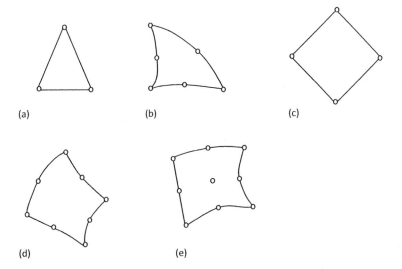

FIGURE 1.11 Some commonly used 2D elements. (a) Three-node triangular, (b) six-node triangular, (c) four-node quadrilateral, (d) eight-node quadrilateral, and (e) nine-node quadrilateral.

natural coordinate space ξ–η, as shown in Figure 1.12, is a good method to compute stiffness matrix of any arbitrary element.

We can relate the x–y coordinates to the ξ and η coordinates as

$$x = N_I x_i + N_{II} x_j + N_{III} x_k \tag{1.29a}$$

$$y = N_I y_i + N_{II} y_j + N_{III} y_k \tag{1.29b}$$

where $(x_i, y_i), (x_j, y_j)$, and (x_k, y_k) are the coordinates of nodes i, j and k (global node numbering) which have element local number I, II, and III. If p is the number of nodes in an element, then it can be defined in terms of local node number l as

$$x = \sum_{l=1}^{p} N_l x_l \tag{1.29c}$$

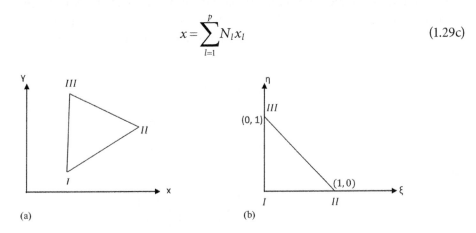

FIGURE 1.12 Mapping of an element into an isoparametric element: (a) Global coordinates. (b) Natural coordinates.

$$y = \sum_{l=1}^{p} N_l y_l \qquad (1.29d)$$

The shape functions are defined as (shown in equation 1.21)

$$N_l = \begin{cases} 1 & \text{at node } l \\ 0 & \text{at all other nodes} \end{cases}$$

and

$$\sum_e N_l = 1$$

We can assume

$$N_I = (1 - \xi - \eta); N_{II} = \xi; N_{III} = \eta \qquad (1.30)$$

For triangular element, the field variable u can be expressed using natural coordinate as

$$u = N_I u_I + N_{II} u_{II} + N_{III} u_{III} \qquad (1.31a)$$

or

$$u = [N]\{u^e\} \qquad (1.31b)$$

In general, it can be expressed as

$$u = \sum_{l=1}^{q} N_l u_l \qquad (1.31c)$$

The order p of geometric function in equation 1.29(c) and (d) and the order q of interpolation function in equation 1.31(c) of triangular (or quadrilateral) element need not be the same. It is called isoparametric element if $p = q$, superparametric element if $p > q$ and subparametric element if $p < q$. Using these relations, we can find the derivative as

$$\frac{\partial u}{\partial \xi} = \frac{\partial u}{\partial x} \cdot \frac{\partial x}{\partial \xi} + \frac{\partial u}{\partial y} \cdot \frac{\partial y}{\partial \xi} \qquad (1.32a)$$

$$\frac{\partial u}{\partial \eta} = \frac{\partial u}{\partial x} \cdot \frac{\partial x}{\partial \eta} + \frac{\partial u}{\partial y} \cdot \frac{\partial y}{\partial \eta} \qquad (1.32b)$$

or

$$\begin{Bmatrix} \dfrac{\partial u}{\partial \xi} \\ \dfrac{\partial u}{\partial \eta} \end{Bmatrix} = \begin{bmatrix} \dfrac{\partial x}{\partial \xi} & \dfrac{\partial y}{\partial \xi} \\ \dfrac{\partial x}{\partial \eta} & \dfrac{\partial y}{\partial \eta} \end{bmatrix} \begin{Bmatrix} \dfrac{\partial u}{\partial x} \\ \dfrac{\partial u}{\partial y} \end{Bmatrix} \qquad (1.33)$$

In the case of three-node triangular isoparametric element where same shape functions are used to represent geometry and interpolation, the Jacobian matrix

$$[J] = \begin{bmatrix} \dfrac{\partial x}{\partial \xi} & \dfrac{\partial y}{\partial \xi} \\ \dfrac{\partial x}{\partial \eta} & \dfrac{\partial y}{\partial \eta} \end{bmatrix} = \begin{bmatrix} x_{ji} & y_{ji} \\ x_{ki} & y_{ki} \end{bmatrix} \tag{1.34}$$

Here $x_{ij} = x_i - x_j$ and $y_{ij} = y_i - y_j$

Therefore

$$\begin{bmatrix} \dfrac{\partial u}{\partial x} & \dfrac{\partial u}{\partial y} \end{bmatrix}^T = [J]^{-1} \cdot \begin{bmatrix} \dfrac{\partial u}{\partial \xi} & \dfrac{\partial u}{\partial \eta} \end{bmatrix}^T \tag{1.35a}$$

or

$$\begin{bmatrix} \dfrac{\partial u}{\partial x} \\ \dfrac{\partial u}{\partial y} \end{bmatrix} = \dfrac{1}{|J|} \begin{bmatrix} y_{ki} & -y_{ji} \\ -x_{ki} & x_{ji} \end{bmatrix} \begin{bmatrix} -u_I + u_{II} \\ -u_I + u_{III} \end{bmatrix} \tag{1.35b}$$

In case of three node triangular element, the determinant of Jacobian matrix $|J| = 2A_e$ (twice of the area of element). Equation (1.35b) can be rearranged as

$$\begin{bmatrix} \dfrac{\partial u}{\partial x} \\ \dfrac{\partial u}{\partial y} \end{bmatrix} = \dfrac{1}{|J|} \begin{bmatrix} y_{jk} & y_{ki} & y_{ij} \\ x_{kj} & x_{ik} & x_{ji} \end{bmatrix} \begin{bmatrix} u_I \\ u_{II} \\ u_{III} \end{bmatrix} \tag{1.36a}$$

$$= [B]\{u^e\} \tag{1.36b}$$

The functional I can be expressed as

$$I = \sum_e t_e \iint_{\Omega_e} \left\{ \dfrac{1}{2} \begin{bmatrix} \dfrac{\partial u^e}{\partial x} & \dfrac{\partial u^e}{\partial y} \end{bmatrix} \begin{bmatrix} \dfrac{\partial u^e}{\partial x} & \dfrac{\partial u^e}{\partial y} \end{bmatrix}^T - uf \right\} d\Omega \tag{1.37a}$$

$$= \sum_e \dfrac{1}{2} \{u^e\}^T \left(t_e \iint_{\Omega_e} [B]^T [B] d\Omega \right) \{u^e\} - \{u^e\}^T t_e \iint_{\Omega_e} [N]^T f d\Omega \tag{1.37b}$$

Here $\{u^e\}$ is the nodal displacement vector and $u^e = u^e(x,y)$ is the interpolation function of the element. Assembly of all elements gives the set of algebraic equations

$$I = \sum_e \frac{1}{2} \{u^e\}^T [k^e]\{u^e\} - \sum_e \{u^e\}^T \{F_i^e\}$$ (1.38a)

$$= \frac{1}{2} \{u\}^T [k]\{u\} - \{u\}^T \{F_i\}$$ (1.38b)

Minimization of functional leads to the equation

$$[k]\{u\} - [F_i] = 0$$

In this case ($d\Omega = dA = dxdy = |J| d\xi\, d\eta = 2 A_e d\xi\, d\eta$), the element stiffness matrix is obtained as

$$\left[k^e \right] = t_e \int_0^1 \int_0^{1-\eta} [B]^T [B] \cdot |J| \cdot d\xi\, d\eta$$ (1.39a)

Since $[B]$ is not a function of ξ and η coordinate,

$$\left[k^e \right] = t_e A_e [B]^T [B]$$ (1.39b)

If we define $\beta_i = x_k - x_j$ and $\gamma_i = y_j - y_k$, then

$$k_{ij}^e = \frac{t_e}{4 A_e} (\beta_i \beta_j + \gamma_i \gamma_j)$$ (1.40)

where i, j, k permute in a natural order and $i \neq j \neq k$.

The force vector for an element can be calculated as

$$\{F_i^e\} = t_e \int_0^1 \int_0^{1-\eta} [N]^T f \cdot |J| \cdot d\xi\, d\eta$$ (1.41)

For $f = f_o$ (a constant)

$$\{F_i^e\} = \frac{t_e A_e f_o}{3} \begin{Bmatrix} 1 \\ 1 \\ 1 \end{Bmatrix}$$ (1.42)

1.5 ELEMENT STIFFNESS MATRIX FOR A BEAM ELEMENT

The beam equation (1.2) can be expressed in the form of potential energy as

$$I = \frac{1}{2} \int_0^L EI \left(\frac{d^2 w}{dx^2} \right)^2 dx - \int_0^L fw\, dx - \sum_i F_i w_i - \sum_j M_j \theta_j$$ (1.43)

which can be expressed after finite element discretization as

$$I = \sum_e \frac{1}{2} \int_{x_I}^{x_{II}} E_e I_e \left(\frac{d^2 w^e}{dx^2} \right)^2 dx - \sum_e \int_{x_I}^{x_{II}} f w^e dx - \sum_i F_i w_i - \sum_j M_j \theta_j \qquad (1.44)$$

Here the force F_i at a node causes transverse deflection w_i and the bending moment M_i at a node causes rotation $\frac{dw}{dx}\big|_i$. Therefore each node has two degrees of freedom specifying transverse deflection and slope. A two node beam element will have four degrees of freedom. Hence we have to assume an interpolation polynomial with four constants (Figure 1.13).

$$w(x) = c_1 + c_2 x + c_3 x^2 + c_4 x^3 \qquad (1.45)$$

The expression can be written in natural coordinate as

$$w(\xi) = a_1 + a_2 \xi + a_3 \xi^2 + a_4 \xi^3 \qquad (1.46)$$

Here

$$x = \frac{(1-\xi)}{2} x_I + \frac{(1+\xi)}{2} x_{II} \qquad (1.47)$$

Therefore

$$dx = \left(\frac{x_{II} - x_I}{2} \right) d\xi \qquad (1.48)$$

and

$$\frac{dw}{d\xi} = \frac{dw}{dx} \cdot \frac{dx}{d\xi} = \frac{l_e}{2} \frac{dw}{dx} \qquad (1.49)$$

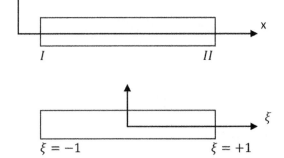

FIGURE 1.13 Transformation of x-coordinate into ξ—a natural coordinate.

The polynomial equation (1.46) can be expressed in terms of deflection and slope as

$$w\left(\text{at } \xi = -1\right) = w_I = a_1 - a_2 + a_3 - a_4 \tag{1.50a}$$

$$\frac{l_e}{2}\frac{dw}{dx}\left(\text{at } \xi = -1\right) = \frac{l_e}{2}\theta_I = a_2 - 2a_3 + 3a_4 \tag{1.50b}$$

$$w\left(\text{at } \xi = +1\right) = w_{II} = a_1 + a_2 + a_3 + a_4 \tag{1.50c}$$

$$\frac{l_e}{2}\frac{dw}{dx}\left(\text{at } \xi = +1\right) = \frac{l_e}{2}\theta_{II} = a_2 + 2a_3 + 3a_4 \tag{1.50d}$$

The constants a_i's can be calculated and the interpolation functions in equation (1.46) can be expressed as

$$w(\xi) = \psi_1 w_I + \psi_2 \theta_I + \psi_3 w_{II} + \psi_4 \theta_{II} \tag{1.51a}$$

$$= \begin{bmatrix} \psi_1 & \psi_2 & \psi_3 & \psi_4 \end{bmatrix} \begin{bmatrix} w_I & \theta_I & w_{II} & \theta_{II} \end{bmatrix}^T \tag{1.51b}$$

$$= [N]\{w^e\} \tag{1.51c}$$

where:

$$\psi_1 = \frac{1}{4}\left(2 - 3\xi + \xi^2\right); \qquad \psi_2 = \frac{l_e}{8}\left(1 - \xi - \xi^2 + \xi^3\right)$$

$$\psi_3 = \frac{1}{4}\left(2 + 3\xi - \xi^3\right); \qquad \psi_4 = \frac{l_e}{8}\left(-1 - \xi + \xi^2 + \xi^3\right)$$

Now

$$\frac{d^2 w}{dx^2} = \left(\frac{4}{l_e^2} \cdot \frac{d^2[N]}{d\xi^2}\right) \cdot \{w^e\} \tag{1.52a}$$

$$= [B]\{w^e\} \tag{1.52b}$$

Therefore stain energy of an element

$$\frac{1}{2}\int_0^l E_e I_e \left(\frac{d^2 w^e}{dx^2}\right)^2 dx = \frac{1}{2}\int_{-1}^{+1} E_e I_e \{w^e\}^T [B]^T [B]\{w^e\}\frac{l_e}{2} d\xi = \frac{1}{2}\{w^e\}^T [k^e]\{w^e\} \tag{1.53}$$

where:

$$\left[k^e\right] = \frac{E_e I_e}{l_e^3} \begin{bmatrix} 12 & 6l_e & -12 & 6l_e \\ 6l_e & 4l_e^2 & -6l_e & 2l_e^2 \\ -12 & -6l_e & 12 & -6l_e \\ 6l_e & 2l_e^2 & -6l_e & 4l_e^2 \end{bmatrix} \tag{1.54}$$

and the load vector

$$\int_{l_e} f w\, dx = \left\{w^e\right\}^T \cdot \frac{l_e}{2} \int_{-1}^{+1} [N]^T f\, d\xi = \left\{w^e\right\}^T \left\{f^e\right\} \tag{1.55}$$

For constant f (upward force)

$$\left\{f^e\right\} = \begin{bmatrix} \dfrac{fl_e}{2} & \dfrac{fl_e^2}{12} & \dfrac{fl_e}{2} & -\dfrac{fl_e}{12} \end{bmatrix} \tag{1.56}$$

In this chapter, we have studied steps of finite element method which we use to solve the problems. We have not presented higher order elements which require numerical integration for finding stiffness matrix. Generally, Gauss-quadrature method is used, which is very well presented in many books. With this knowledge, students can write the basic computer program, but they have to learn the methods of storing stiffness matrix and its various solution procedures for efficient programming.

The results of the finite element method are reliable in a large class of engineering problems but not all the problems. There are some issues on its mathematics and applications which can be discussed after basic knowledge on spaces (Chapter 3) and operators (Chapter 4). We present them in Chapter 5.

REFERENCES

1. T.R. Chandrupatla and A.D. Belegundu (2006), *Introduction to Finite Element in Engineering*, (3rd ed.), Pearson Education, New Delhi, India.
2. D.V. Hutton (2004), *Fundamentals of Finite Element Analysis*, McGraw-Hill Companies, New York.
3. S.S. Rao (2004), *The Finite Element Method in Engineering*, (4th ed.), Elsevier Inc, Burlington, MA.
4. J.N. Reddy (2005), *An Introduction to the Finite Element Method*, (3rd ed.), McGraw-Hill Education, New York.
5. S. A. Ragab and H.E. Fayed (2018), *Introduction to Finite Element Analysis for Engineers*, CRC Press, Boca Raton, FL.

CHAPTER 2

Introduction to Wavelets for Solution of Differential Equations

I N THIS CHAPTER, SOME BASIC approaches of wavelets for the solution of partial differential equations (PDEs) are presented. The concept of spaces is necessary for proper understanding of wavelets but for simplicity, the strenuous concept of functional analysis is avoided in the beginning. Moreover, this process will generate curiosity and interest on the abstract concepts of functional analysis which are presented in Chapters 3 and 4.

To introduce some fundamental wavelet based techniques for solution of PDEs, we will present here only essential concepts of wavelets in a simple way. Wavelets are most widely used in signal and image processing because of their efficient low and high pass filtering algorithms. Here we will not use wavelets as high and low pass filters but discuss how to solve differential equations. Wavelets offer many interesting techniques to solve various types of differential equations because of their multiresolution character, spectral nature, high and low frequency components, compact support, orthogonal and vanishing moment properties. But the methods are not as straightforward as FEM because satisfying boundary and initial conditions of engineering problems and solution of two and three dimensional complex domain problems are not easy. Students can understand wavelets in a formal way after reading Chapters 3 and 4 on functional analysis.

2.1 WAVELET BASIS FUNCTIONS

To solve a differential equation

$$Au = f \tag{2.1}$$

where A is a differential operator, we approximate the field variable u in terms of the basis[*] functions φ_i as

$$u \approx \sum_i c_i \varphi_i \qquad (2.2)$$

where c_i are constants. Fourier, Legendre and Chebyshev, etc. are well known basis functions. In wavelet based methods, it can be expressed as:

$$P_j(u) = \sum_i c_i^j \varphi_i^j \qquad (2.3a)$$

where φ_i^j is called scaling function. $P_j(u)$ represents the projection of function u onto the subspace[*] spanned by the basis function. Figure 2.1a shows projection using the simplest scaling function developed by Haar in 1909.

Mathematica programs presented at the end of the chapter show how the exact function u is projected at some scale j. Similarly, the projection will be refined at $j+1$ level and the number of discrete points is doubled. The approximation at one level higher scaling function is shown in Figure 2.1b. For engineering purpose, the level $j+1$ is used for 2^{j+1} discrete points. This projection is expressed as

$$P_{j+1}(u) = \sum_i c_i^{j+1} \varphi_i^{j+1} \qquad (2.3b)$$

The difference between the two consecutive projections is approximated by using basis functions ψ_i^j called wavelets. This can be written as

$$Q_j(u) = \sum_i d_i^j \psi_i^j \qquad (2.4)$$

and it is shown in Figure 2.1c (generated by Mathematica program). Therefore, the approximation of u is improved when the two projections $P_j(u)$ and $Q_j(u)$ are combined

$$P_j(u) + Q_j(u) = P_{j+1}(u). \qquad (2.5a)$$

In other words, the approximation of field variable can be refined using wavelets

$$u \approx \sum_i c_i^j \varphi_i^j + \sum_i d_i^j \psi_i^j = \sum_i c_i^{j+1} \varphi_i^{j+1} \qquad (2.5b)$$

It can be interpreted from equation (2.5) that the basis function at a lower level j can be expressed in terms of the basis functions of higher level $j+1$ and vice versa.

---

[*] Basis function and subspaces are defined in Chapter 3.

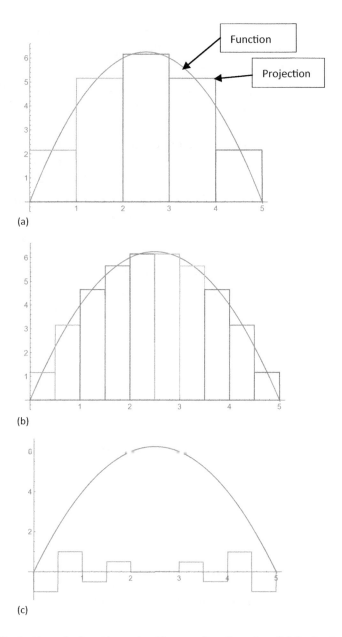

FIGURE 2.1   (a) Projection of a function using Haar scaling function. (b) Projection using one level higher scaling function. (c) Projection of Haar wavelet.

$$\varphi_k^j = \sum_i h_i \varphi_i^{j+1} \tag{2.6a}$$

and

$$\psi_k^j = \sum_i g_i \varphi_i^{j+1} \tag{2.6b}$$

This is known as *refinement relations*. The coefficients h_i and g_i are known as *low and high pass filter coefficients* because they split a signal into high and low frequency signals. For more details, we suggest students read some books of signal processing. In this chapter, this knowledge is not required. These simple looking filter coefficients are the most interesting feature in the mathematics of wavelets which we will understand in Chapter 6.

In Figure 2.1a–c, we demonstrated the projection process using Haar scaling function and wavelets. The process and the figures will be very similar for other wavelets such as Daubechies, etc. Figure 2.2 shows Haar scaling functions which is a box function between 0 and 1. As shown in the figure, all other scaling functions are generated by translates of one function. Similarly, Figure 2.3 shows Haar wavelets which are also translates of one wavelet function. These basis functions have compact support, which means function is nonzero at some small interval.

Figure 2.3 shows the scaling functions at the next higher level $j+1$. Compare Figures 2.2 and 2.3 to see the dilation, which is a very important property of scaling function and wavelets. The translation property, shown in Figure 2.2, can be shown at any level. In case

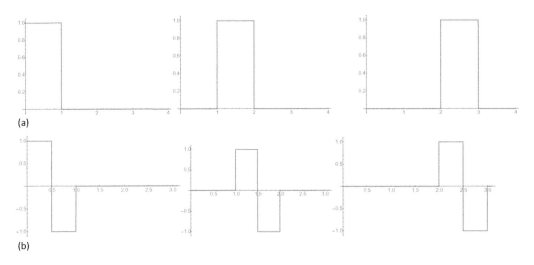

FIGURE 2.2   (a) Haar scaling function and its translates at some scale j: $\varphi_0^j, \varphi_1^j, \varphi_2^j$. (b) Haar wavelets and its translates at some scale j: $\psi_0^j, \psi_1^j, \psi_2^j$.

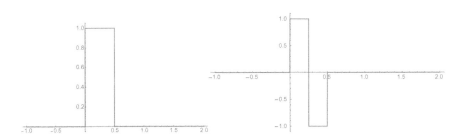

FIGURE 2.3   At next higher scale $j+1$: Haar scaling function φ_0^{j+1} and wavelet ψ_0^{j+1}.

of Haar scaling functions and wavelets, the refinement relations equation (2.6) can be expressed as

$$\varphi_i^j = \varphi_i^{j+1} + \varphi_{i+1}^{j+1} \tag{2.7a}$$

and

$$\psi_i^j = \varphi_i^{j+1} - \varphi_{i+1}^{j+1} \tag{2.7b}$$

It is essential to have basis functions φ_i^j and ψ_i^j in terms of basis functions φ_i^{j+1} as expressed in the case of Haar. The students can easily validate the Haar refinement relations. The basis function shown in Figure 2.3 along with the filter coefficients $h = \{1,1\}$ and $g = \{1,-1\}$ (equation (2.7)) can be used to generate the basis function shown in Figure 2.2.

It is easy to verify the orthogonal property of Haar scaling function and wavelet, i.e.,

$$\int_{-\infty}^{\infty} \varphi_k^j \varphi_m^j dx = \begin{cases} 1 & \text{for } k = m \\ 0 & \text{otherwise} \end{cases} \tag{2.8a}$$

$$\int_{-\infty}^{\infty} \psi_k^j \psi_m^j dx = \begin{cases} 1 & \text{for } k = m \\ 0 & \text{otherwise} \end{cases} \tag{2.8b}$$

and

$$\int_{-\infty}^{\infty} \varphi_k^j \psi_m^j dx = 0 \text{ for all } k \text{ and } m \tag{2.8c}$$

Due to orthogonal property of scaling functions and wavelets, the coefficients c_i^j, d_i^j etc. of equation (2.5) can be obtained as

$$c_i^j = \int_{-\infty}^{\infty} u \varphi_i^j dx \tag{2.9a}$$

and

$$d_i^j = \int_{-\infty}^{\infty} u \psi_i^j dx \tag{2.9b}$$

The orthogonal property is very useful, but all scaling functions and wavelets are not orthogonal.

It is interesting to note that the unknown coefficients can be obtained by using the filter coefficients h_i and g_i. The proof is presented in Chapter 6. For example, in the case of Haar, these coefficients can be obtained as

$$c_i^j \approx \frac{1}{2} u^j[i] + \frac{1}{2} u^j[i+1] \tag{2.10a}$$

and

$$d_i^j \approx \frac{1}{2} u^j[i] - \frac{1}{2} u^j[i+1] \tag{2.10b}$$

The filter coefficients $h = \{1,1\}$ and $g = \{1,-1\}$ are used in equation (2.7) whereas $h = \{\frac{1}{2}, \frac{1}{2}\}$ and $g = \{\frac{1}{2}, -\frac{1}{2}\}$ are used in the preceding equation. For signal processing applications, a normalized filter coefficient matrix contains $h = \{\frac{1}{\sqrt{2}}, \frac{1}{\sqrt{2}}\}$ and $g = \{\frac{1}{\sqrt{2}}, -\frac{1}{\sqrt{2}}\}$. Such practice is common in all the wavelets.

2.2 WAVELET-GALERKIN METHOD

Wavelet-Galerkin method is very similar to the finite element method and appears to be more useful than the collocation method for engineering problems; therefore we are considering only the former method. Substitution of equation (2.5) as a trial function $\tilde{u}$ for the solution of differential equation (2.1) leads to some residual

$$R = A\tilde{u} - f \tag{2.11}$$

Here we are using only one scale, so the scale notation j (of φ_i^j) is dropped and we define $\tilde{u} = P_j(u)$. The unknown coefficients c_i and d_i are obtained by imposing the conditions that the residual equation (2.11) be orthogonal to the test functions

$$\int_\Omega v(A\tilde{u} - f) d\Omega = 0 \tag{2.12a}$$

In 1917, Russian engineer V.I. Galerkin suggested that the test function v is of the same form as $\tilde{u}$. For trial function equation (2.2), the equation (2.12a) can be expressed as

$$\int_\Omega \varphi_i \left(A \sum_{l=1}^n c_l \varphi_l - f \right) d\Omega = 0 \qquad \text{for } i = 1, \dots n \tag{2.12b}$$

Let us define

$$k_{il} = \int_\Omega \varphi_i A \varphi_l d\Omega \tag{2.13a}$$

and

$$f_i = \int_{\Omega} \varphi_i f d\Omega \qquad (2.13b)$$

Initially, the students find difficulty in understanding the process of differentiation and integration of Daubechies wavelets [1]; therefore without discussing the process, we used numerical values of integrals directly in the program. The process of differentiation and integration is explained in Appendix D.

The set of equations (2.12b) can be expressed as

$$\sum_{l=1}^{n} k_{il} c_l = f_i \qquad \text{for } i = 1, \dots n \qquad (2.14)$$

Initially, we consider only one level of refinement of multiscale approach

$$u \approx \sum_i c_i^0 \varphi_i^0 + \sum_l d_i^j \psi_i^j$$

i.e., trial function equation (2.5) and test functions φ_i^j and ψ_i^j, (for convenience, superscript j is dropped) the equation (2.12a) can be written as

$$\int_{\Omega} \varphi_i \left(A \left[\sum_l c_l \varphi_l + \sum_l d_l \psi_l \right] - f \right) d\Omega = 0 \qquad \text{for } i = 1, \dots .n \qquad (2.15a)$$

and

$$\int_{\Omega} \psi_i \left(A \left[\sum_l c_l \varphi_l + \sum_l d_l \psi_l \right] - f \right) d\Omega = 0 \qquad \text{for } i = 1, \dots n \qquad (2.15b)$$

which can be expressed in the matrix form as

$$\begin{bmatrix} k_{11} & . & . & k_{1n} & \alpha_{11} & . & . & \alpha_{1m} \\ . & . & . & . & . & . & . & . \\ . & . & . & . & . & . & . & . \\ k_{n1} & . & . & k_{nn} & \alpha_{n1} & . & . & \alpha_{nm} \\ \beta_{11} & . & . & \beta_{1n} & \gamma_{11} & . & . & \gamma_{1m} \\ . & . & . & . & . & . & . & . \\ . & . & . & . & . & . & . & . \\ \beta_{m1} & . & . & \beta_{mn} & \gamma_{m1} & . & . & \gamma_{mm} \end{bmatrix} \begin{bmatrix} c_1 \\ . \\ . \\ c_n \\ d_1 \\ . \\ , \\ d_m \end{bmatrix} = \begin{bmatrix} f_1 \\ . \\ . \\ f_n \\ g_1 \\ . \\ . \\ g_m \end{bmatrix} \qquad (2.16a)$$

where:

$$\alpha_{il} = \int_{\Omega} \varphi_i A\psi_l d\Omega, \quad \beta_{il} = \int_{\Omega} \psi_i A\varphi_l d\Omega, \quad \gamma_{il} = \int_{\Omega} \psi_i A\psi_l d\Omega \qquad (2.16b)$$

and

$$g_i = \int_{\Omega} \psi_i f d\Omega \qquad (2.16c)$$

2.3 DAUBECHIES WAVELETS FOR BOUNDARY AND INITIAL VALUE PROBLEMS

We need a basis function whose derivative must be non-zero and finite; therefore, Haar scaling function and wavelets cannot be used for the solution of differential equations. In 1988, Ingrid Daubechies proposed a method to develop wavelets with compact support which are orthogonal and differentiable, often represented as D2, D4, D6,… DN. Here N represents number of coefficients which have N/2 vanishing moments. The details are presented in Chapter 6. These wavelets are also represented as db1, db2, db3, etc. In this chapter, we will use Daubechies wavelets D6, which are often used to solve PDEs. An introduction of Daubechies wavelets is presented in Appendix D, but the features required for solution of PDEs are presented here.

We cannot use Daubechies scaling functions and wavelets as we use the polynomials in FEM. To understand it, let us see how a Daubechies scaling function is generated. Figure 2.4 shows the points of scaling function at 1st, 2nd and 3rd steps generated with the help of a MATLAB program (Program 2.1). A continuous curve can be generated with the increase of the number of steps. It is interesting to investigate how the program utilizes filter coefficients h. It will not be difficult to understand that the differentiation and integration of such basis function will not be an easy task whose interpolating points are generated in such an iterative way. In this chapter, we will not discuss how the basis functions are generated, differentiated and integrated, but we will use the results. Readers can refer to Appendix D.

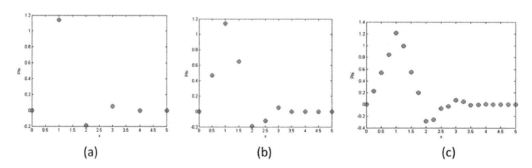

(a)   (b)   (c)

FIGURE 2.4   Generation of scaling function at (a) first step, (b) second step, (c) third step.

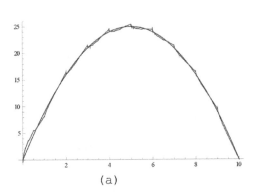
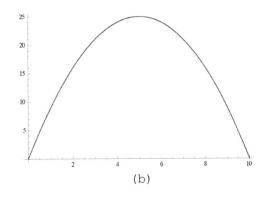

(a)

(b)

FIGURE 2.5 Daubechies scaling function to generate $-x^2 + 10x$: (a) D4 and (b) D6.

The scaling functions of Daubechies wavelets DN can generate a polynomial of degree $p = (N/2) - 1$. A polynomial generated by D4 and D6 scaling functions using the Mathematica program is shown in Figure 2.5. The students should try other scaling functions.

To solve a differential equation, we have to calculate coefficients k_{il}, α_{il}, β_{il}, γ_{il}, f_i, g_i of the equation (2.15). Defining the following notation for derivative

$$\varphi_i^d = \frac{\partial^d \varphi}{\partial x^d} \tag{2.17}$$

The coefficients of the matrix will have integrals of the form

$$\Omega^{d_1 d_2}[i - l] = \int_{-\infty}^{\infty} \varphi_i^{d_1}\, \varphi_l^{d_2}\, dx \tag{2.18}$$

which is called two term connection coefficients. The differentiations, integrations and multiplications are a tricky task. Fortunately, some efficient algorithms are developed [1] for this purpose. In this chapter, either the results of these algorithms will be used or we will use Mathematica/MATLAB functions.

It can be noted in Figure 2.6 that D6 scaling functions phiD6(x) (generated with the help of Mathematica) is non-zero for $0 \leq x \leq 5$, and in this support, only a few scaling functions represented as phiD6($x + 4$) to phiD6($x - 4$) are non-zero. Therefore in this range, the connection coefficients equation (2.18) or the product of phiD6(x) with the basis function phiD6($x + 4$) to phiD6($x - 4$) will be non-zero. In general, connection coefficients (equation 2.18) of DN will be non-zero if $|i - l| \leq N - 2$. The stiffness matrix will be sparse due to a large number of zero connection coefficients.

Due to overlapping of scaling functions, shown in Figure 2.6, application of boundary conditions and initial conditions is not as simple as FEM. The method proposed by Amaratunga and Williams [2–6] will be used in the examples.

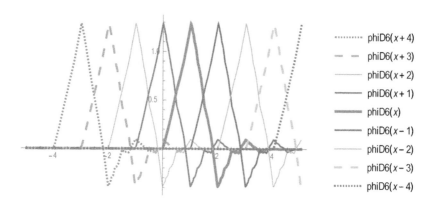

FIGURE 2.6   D6 scaling functions at $x = i$ (some integer) and the neighboring scaling functions.

Example 2.1: Boundary Value Problem

Consider a differential equation

$$\frac{d^2 u}{dx^2} = f$$

With some Dirichlet boundary conditions

$$u = 2 \text{ at } x = 0.2 \text{ and } u = 3 \text{ at } x = 0.7$$

This problem is solved in two steps; first we solve it as a periodic problem with period d. Let v equal the solution of periodic problem. Next we use a green function to satisfy the boundary conditions. Let w equal the solution to this part [7]. The solution of this problem will be

$$u = v + w$$

Part 1. Periodic Problem:
   We can define the periodic problem as

$$\frac{d^2 v}{dx^2} = f$$

For the periodic problem, $v(0) = v(d)$ and $f(0) = f(d)$. For convenience, we assume $d = 1$. The function, which is defined for $0.2 \leq x \leq 0.7$, can be extended from 0 to d by adding zeros.

   Similar to equation (2.2), $v(x)$ and $f(x)$ can be approximated at scale j by scaling function

$$v(x) = \sum v_k 2^{j/2} \varphi\left(2^j x - k\right)$$

$$f(x) = \sum f_k 2^{j/2} \varphi\left(2^j x - k\right)$$

where v_k and f_k are coefficients.

If we substitute it in the differential equation, then

$$\frac{d^2}{dx^2}\left[\sum v_k 2^{j/2}\varphi\left(2^j x-k\right)\right]=\sum f_k 2^{j/2}\varphi\left(2^j x-k\right)$$

Let $y=2^j x$

$$2^{2j}\frac{d^2}{dy^2}\left[\sum v_k\varphi\left(y-k\right)\right]=\sum f_k\varphi\left(y-k\right)$$

The inner product of both sides with $\varphi(y-l)$ yields ($\langle u,v\rangle=\int u\cdot v$ is called the inner product. It is discussed in Chapter 3.)

$$2^{2j}\sum v_k\int\varphi''\left(y-k\right)\varphi\left(y-l\right)dy=\sum f_k\int\varphi\left(y-k\right)\varphi\left(y-l\right)dy$$

Due to orthogonality

$$\int\varphi\left(y-k\right)\varphi\left(y-l\right)dy=\delta_{lk}$$

If we take Daubechies scaling function of order N, then the integral $\int\varphi''\left(y-k\right)\varphi\left(y-l\right)dy$ will be non-zero for $k=l\pm(N-2)$, and for all other k's the two scaling functions will not have any common part of support. Therefore the equation can be written as

$$2^{2j}\sum_{k=l-N+2}^{l+N-2}v_k\Omega_{l-k}^2=f_l \tag{2.19}$$

where Ω_{l-k}^2 is the connection coefficient.

$$\Omega_{l-k}^2=\int\varphi''\left(y-k\right)\varphi\left(y-l\right)dy$$

For convenience, we write Ω_{l-k}^2 as $\Omega[l-k]$. In the present example, we are using Daubechies scaling function of order 6. For $l=0$, the equation can be expanded as

$$2^{2j}\left[v_{-4}\Omega[4]+v_{-3}\Omega[3]+v_{-2}\Omega[2]+v_{-1}\Omega[1]+v_0\Omega[0]\right.$$

$$\left.+v_1\Omega[-1]+v_2\Omega[-2]+v_3\Omega[-3]+v_4\Omega[-4]\right]=f_0$$

Since it is periodic,

$$v_{-4}=v_{n-4};\ v_{-3}=v_{n-3};\ v_{-2}=v_{n-2};\ v_{-1}=v_{n-1}$$

Similarly, at the end $l = n-1$ $(n = 2^j)$, equation (2.19) can be expressed as

$$2^{2j}\left[v_{n-5}\Omega[4] + v_{n-4}\Omega[3] + v_{n-3}\Omega[2] + v_{n-2}\Omega[1] + v_{n-1}\Omega[0] \right.$$

$$\left. + v_n\Omega[-1] + v_{n+1}\Omega[-2] + v_{n+2}\Omega[-3] + v_{n+3}\Omega[-4] \right] = f_{n-1}$$

Due to periodicity; $v_n = v_0$; $v_{n+1} = v_1$; $v_{n+2} = v_2$; $v_{n+3} = v_3$

All other equations for $l = 1$ to $n-2$ can be obtained and expressed in the matrix form as

$$[T][v] = [f]$$

We can find $[v] = [T]^{-1}[f]$, but Amartunga [7] suggested a more efficient method for the circulent matrix $[T]$. Let $[T_\Omega]$ be the first column of $[T]$. If $\left[\hat{T}_\Omega\right]$ and $\left[\hat{f}\right]$ are Fourier Transforms of $[T_\Omega]$ and $[f]$, respectively, then $\left[\hat{v}\right]$, the Fourier Transform of $[v]$, can be obtained as

$$\left[\hat{v}\right] = \left[\hat{f}\right] / \left[\hat{T}_\Omega\right]$$

The inverse Fourier Transform gives the solution $[v]$. But this solution does not satisfy the boundary conditions, i.e., $v_{at\,0.2} \neq u_{at\,0.2}$ and $v_{at\,0.7} \neq u_{at\,0.7}$.

Therefore, a Green function based method is used to satisfy the boundary conditions.

Part 2. Incorporation of Boundary Conditions:

Incorporation of Boundary Conditions:

Now we have to find w so that $u_a = v_a + w_a$ and $u_b = v_b + w_b$ for given $u_a = u(0.2) = 2$ and $u_b = u(0.7) = 3$. The values of v_a and v_b are calculated in Part 1, so we get $w_a = u_a - v_a$ and $w_b = u_b - v_b$.

If $G(y)$ is the green function of the differential equation,

$$\frac{d^2G}{dx^2} = \delta(x)$$

Similar to Part 1, the Fourier transform based method can be used to find Green function $G(x)$, which is the deflection due to unit concentrated force. Let X_a and X_b be the forces applied at $x = a$ and $x = b$ to generate deflections w_a and w_b, then deflection at x

$$w(x) = G(x) \cdot X(x) = X_a G(x-a) + X_b G(x-b)$$

From this equation we have

$$w(a) = X_a G(0) + X_b G(a-b)$$

$$w(b) = X_b G(b-a) + X_b G(0)$$

The values of X_a and X_b can be calculated by using known Green function calculated in the previous step, and for known $w(a) = w_a$ and $w(b) = w_b$. The two concentrated forces X_a and X_b, acting at $x = a$ and $x = b$, cause deflection w_a and w_b therefore satisfy the boundary conditions. The solution $w(x)$ due to force $X(x) = X_a \delta(x-a) + X_b \delta(x-b)$ can be obtained by solving

$$\frac{d^2 w}{dx^2} = X$$

The method used in Part 1 can be applied to find the solution $w(x)$. Now $u = v + w$ will be the solution which will satisfy the boundary conditions also. This boundary value problem is solved by a MATLAB program given at the end of the chapter.

Example 2.2: Initial Value Problem

This wavelet based method for the initial value problems is developed by Amaratunga [5]. The method is extensively used for wave propagation [8]. Consider a second order ordinary differential equation

$$\frac{d^2 u}{dt^2} = f(t)$$

with initial conditions

$$u(t_0) = u_0 \text{ and } \left. \frac{du}{dt} \right|_{t_0} = \dot{u}(t_0) = u_{t_0}$$

Similar to the previous problem, we express $u(t)$ and $f(t)$ in terms of scaling functions as

$$u(t) = \sum c_k \varphi(2^m t - k)$$

$$f(t) = \sum g_k \varphi(2^m t - k)$$

Let us define a new time scale $\tau = 2^m t$. The Wavelet-Galerkin scheme gives

$$\int \varphi(\tau - j) \left[\ddot{u}(\tau) - f(\tau) \right] d\tau = 0$$

Similar to previous problem, substituting

$$\frac{d^2 u}{dt^2} = \ddot{u} = 2^{2m} \sum_{k=k_0}^{k_1-1} c_k \ddot{\varphi}(\tau - k)$$

and defining $\int \varphi(\tau - j) \ddot{\varphi}(\tau - k) = \Omega[j-k]$.

Here Ω is the connection coefficient and k_0 is the starting time which we assumed as 4 in the program. The method to calculate the connection coefficient is explained in [1]. The connection coefficients are used and listed in the program. The inner product $\langle \varphi(\tau-j)\cdot\ddot{\varphi}(\tau-k)\rangle$ will be non-zero between $k=j-N+2$ to $j+N-2$, therefore

$$2^{2m}\sum_{k=j-N+2}^{j+N-2} c_k\Omega_{j-k} = g_j \tag{2.20}$$

Here $\int\varphi(\tau-j)f(\tau)=g_j$.

In the case of an initial value problem, we express the unknown variables c_k in terms of known variables c_{k-i}. Here we generate a polynomial using known variables and then find the unknown c_k using a polynomial. If a polynomial passes through a point $\tau=j$, then we can express a polynomial of order N as

$$P_N = \lambda_0 + \lambda_1(\tau-j) + \lambda_2(\tau-j)^2 \ldots + \lambda_N(\tau-j)^{N-1}$$

A Daubechies scaling function which has N coefficients can exactly represent a polynomial of order $p-1=\frac{N}{2}-1$; therefore, a polynomial about the point $\tau=j+1$ can be expressed as

$$\sum_{k=k_0-N+2}^{k_1-1} c_k\varphi(\tau-k) = \sum_{r=0}^{p}\lambda_r(\tau-j-1)^r$$

The inner product of both sides with $\varphi(\tau-l)$ gives

$$\left\langle \sum_{k=k_0-N+2}^{k_1-1} c_k\varphi(\tau-k),\varphi(\tau-l)\right\rangle = \left\langle \sum_{r=0}^{p}\lambda_r(\tau-j-1)^r,\varphi(\tau-l)\right\rangle$$

Due to orthogonality of Daubechies scaling functions

$$\int\varphi(\tau-k)\cdot\varphi(t-l) = \begin{cases}1 \text{ for } k=l \\ 0 \text{ otherwise}\end{cases}$$

Therefore,

$$c_k = \sum_{r=0}^{p}\lambda_r\int(\tau-j-1)^r\varphi(\tau-k)d\tau = \sum_{r=0}^{p}\lambda_r\mu_{k-j-1}^r \tag{2.21}$$

For $p = 2$ $(N = 6)$

$$k = j: \quad c_j = \lambda_0 \mu_{-1}^0 + \lambda_1 \mu_{-1}^1 + \lambda_2 \mu_{-1}^2$$

$$k = j - 1: \quad c_{j-1} = \lambda_0 \mu_{-2}^0 + \lambda_1 \mu_{-2}^1 + \lambda_2 \mu_{-2}^2$$

$$k = j - 2: \quad c_{j-2} = \lambda_0 \mu_{-3}^0 + \lambda_1 \mu_{-3}^1 + \lambda_2 \mu_{-3}^2$$

Or

$$\begin{bmatrix} \mu_{-1}^0 & \mu_{-1}^1 & \mu_{-1}^2 \\ \mu_{-2}^0 & \mu_{-2}^1 & \mu_{-2}^2 \\ \mu_{-3}^0 & \mu_{-3}^1 & \mu_{-3}^2 \end{bmatrix} \begin{bmatrix} \lambda_0 \\ \lambda_1 \\ \lambda_2 \end{bmatrix} = \begin{bmatrix} c_j \\ c_{j-1} \\ c_{j-2} \end{bmatrix}.$$

It means that at $\tau = j + 1$, we are calculating polynomial coefficients $(\lambda_0, \lambda_1, \lambda_2)$ using previous three (known) scaling coefficients c_j. The inverse of the matrix gives the unknown polynomial coefficients as

$$\lambda_0 = \xi_{0,0} c_{j-2} + \xi_{0,1} c_{j-1} + \xi_{0,2} c_j$$

$$\lambda_1 = \xi_{1,0} c_{j-2} + \xi_{1,1} c_{j-1} + \xi_{1,2} c_j$$

$$\lambda_2 = \xi_{2,0} c_{j-2} + \xi_{2,1} c_{j-1} + \xi_{2,2} c_j$$

Using equation (2.21), we can express the unknown scaling function coefficients as

$$c_k = \lambda_0 \mu_{k-j-1}^0 + \lambda_1 \mu_{k-j-1}^1 + \lambda_2 \mu_{k-j-1}^2$$

$$= [\xi_{0,0} \mu_{k-j-1}^0 + \xi_{1,0} \mu_{k-j-1}^1 + \xi_{2,0} \mu_{k-j-1}^2] c_{j-2}$$

$$+ [\xi_{0,1} \mu_{k-j-1}^0 + \xi_{1,1} \mu_{k-j-1}^1 + \xi_{2,1} \mu_{k-j-1}^2] c_{j-1}$$

$$+ [\xi_{0,2} \mu_{k-j-1}^0 + \xi_{1,2} \mu_{k-j-1}^1 + \xi_{2,2} \mu_{k-j-1}^2] c_j$$

or $c_k = \upsilon_{k-j-1,0} c_{j-2} + \upsilon_{k-j-1,1} c_{j-1} + \upsilon_{k-j-1,2} c_j$

The unknown scaling coefficients $c_{j+1}, c_{j+2}, c_{j+3}$ and c_{j+4} are obtained as

$$c_{j+1} = \upsilon_{0,0} c_{j-2} + \upsilon_{0,1} c_{j-1} + \upsilon_{0,2} c_j$$

$$c_{j+2} = \upsilon_{1,0} c_{j-2} + \upsilon_{1,1} c_{j-1} + \upsilon_{1,2} c_j$$

$$c_{j+3} = \upsilon_{2,0}c_{j-2} + \upsilon_{2,1}c_{j-1} + \upsilon_{2,2}c_j$$

$$c_{j+4} = \upsilon_{3,0}c_{j-2} + \upsilon_{3,1}c_{j-1} + \upsilon_{3,2}c_j$$

For $N=6$, we can write equation (2.20) as

$$2^{2m}\left[c_{j-4}\Omega_4 + c_{j-3}\Omega_3 + c_{j-2}\Omega_2 + c_{j-1}\Omega_1 + c_j\Omega_0 + c_{j+1}\Omega_{-1} + c_{j+2}\Omega_{-2} + c_{j+3}\Omega_{-3} + c_{j+4}\Omega_{-4}\right] = g_j$$

Substituting values of $c_{j+1}, c_{j+2}, c_{j+3}$ and c_{j+4} in the preceding equation, we get

$$2^{2m}\left[c_{j-4}\Omega_4 + c_{j-3}\Omega_3 + c_{j-2}\left(\Omega_2 + \Theta_1\right) + c_{j-1}\left(\Omega_1 + \Theta_2\right) + c_j\left(\Omega_0 + \Theta_3\right)\right] = g_j$$

where $\Theta_1 = \upsilon_{0,0}\Omega_{-1} + \upsilon_{1,0}\Omega_{-2} + \upsilon_{2,0}\Omega_{-3} + \upsilon_{3,0}\Omega_{-4}$

$$\Theta_2 = \upsilon_{0,1}\Omega_{-1} + \upsilon_{1,1}\Omega_{-2} + \upsilon_{2,1}\Omega_{-3} + \upsilon_{3,1}\Omega_{-4}$$

$$\Theta_3 = \upsilon_{0,2}\Omega_{-1} + \upsilon_{1,2}\Omega_{-2} + \upsilon_{2,2}\Omega_{-3} + \upsilon_{3,2}\Omega_{-4}$$

The unknown c_j can be determined by using four previous known c_{j-i}. A MATLAB program for the initial value problem is included at end of the chapter. The single order differential equation can be solved in a similar way.

The moments of the scaling function are calculated using the following recursion formulae:

$$\mu_0^0 = \int_{-\infty}^{\infty} \varphi(\tau)\,d\tau = 1$$

$$\mu_0^l = \frac{1}{2(2^l-1)}\sum_{i=0}^{l-1}\binom{l}{i}\left(\sum_{k=0}^{N-1}a_k k^{l-i}\right)\mu_0^i$$

$$\mu_k^l = \sum_{i=0}^{l}\binom{l}{i}k^{l-i}\mu_0^i$$

where a_k are filter coefficients in the dilation equation

$$\varphi(t) = \sum_{k=0}^{N-1} a_k \varphi(2t - k)$$

Initial Conditions:

To find the first four values of c_j, one-step finite difference method can be used. As an example, for first order differential equation $u_t(t) = f(u, t)$ with initial conditions $u(t=0) = u_0$. Our objective is to find the first four approximate values of $u(t_n)$ for $t_n = n\Delta t$.

We can use the simplest one-step finite difference method defined as $u(t_{n+1}) = u(t_n) + \Delta t \cdot f(u(t_n), t_n)$. In the present problem, $f = t^2 - t$ with initial conditions $u(0) = 5$. We assume $\Delta t = 1/(2^6)$, then

$$u(t_1) = u(t_0) + \Delta t \cdot f(t_0) = 5$$

$$u(t_2) = u(t_1) + \Delta t \cdot f(t_1) = 4.9998$$

$$u(t_3) = u(t_2) + \Delta t \cdot f(t_2) = 4.9993$$

Similarly for the second order equation $u_{tt}(t) = f(u_t, u, t)$, we can find the first four values of $u(t_n)$ using the formulae:

$$u_{tt}(t_n) = f\left(u_t(t_n), u(t_n), t_n\right)$$

$$u(t_{n+1}) = u(t_n) + \Delta t \cdot u_t(t_n) + \frac{\Delta t^2}{2} \cdot u_{tt}(t_n)$$

$$u_t(t_{n+1}) = u_t(t_n) + \Delta t \cdot u_{tt}(t_n)$$

For $f = t^2 - t$ with initial conditions $u(0) = 1$ and $u_t(0) = 0.2$ and $\Delta t = 1/(2^6)$ we get

$$u(t_0) = 1.0$$

$$u(t_1) = 1.0031$$

$$u(t_2) = 1.0062$$

$$u(t_3) = 1.0094$$

MATLAB Program for Figure 2.1:

```
phihaar[x_] = WaveletPhi[HaarWavelet[], x];

u = -x^2 + 5 x;

c = {};

For[i = 0; c1, i < 11, i++,
  c1 = NIntegrate[u * phihaar[x - i], {x, 0, 10}, Method → "GaussKronrodRule",
    AccuracyGoal → 6, MaxRecursion → 20, WorkingPrecision → 10]; AppendTo[c, c1];]

PjU = c[[1]] * phihaar[x] + c[[2]] * phihaar[x - 1] +
  c[[3]] * phihaar[x - 2] + c[[4]] * phihaar[x - 3] + c[[5]] * phihaar[x - 4];

Plot[{u, PjU}, {x, -0.01, 5.01}, Exclusions → None, PlotRange → All];

c = {};

For[i = 0; c1, i < 11, i++,
  c1 = NIntegrate[2 * u * phihaar[2 x - i], {x, 0, 10}, Method → "GaussKronrodRule",
    AccuracyGoal → 6, MaxRecursion → 20, WorkingPrecision → 10]; AppendTo[c, c1];]

Pj1U = c[[1]] * phihaar[2 x] + c[[2]] * phihaar[2 x - 1] +
  c[[3]] * phihaar[2 x - 2] + c[[4]] * phihaar[2 x - 3] + c[[5]] * phihaar[2 x - 4] +
  c[[6]] * phihaar[2 x - 5] + c[[7]] * phihaar[2 x - 6] + c[[8]] * phihaar[2 x - 7] +
  c[[9]] * phihaar[2 x - 8] + c[[10]] * phihaar[2 x - 9];

Plot[{u, Pj1U}, {x, -0.01, 5.01}, Exclusions → None, PlotRange → All];

psihaar[x_] = WaveletPsi[DaubechiesWavelet[1], x];

c = {};

For[i = 0; c1, i < 5, i++,
  c1 = NIntegrate[u * psihaar[x - i], {x, 0, 5}, Method → "GaussKronrodRule",
    AccuracyGoal → 6, MaxRecursion → 20, WorkingPrecision → 10]; AppendTo[c, c1];]

QjU = c[[1]] * psihaar[x] + c[[2]] * psihaar[x - 1] +
  c[[3]] * psihaar[x - 2] + c[[4]] * psihaar[x - 3] + c[[5]] * psihaar[x - 4];

Plot[{u, QjU}, {x, -0.01, 5.01}, Exclusions → None, PlotRange → All];
```

MATLAB Program for Figures 2.2 and 2.3:

```
u = WaveletPhi[HaarWavelet[], x];

v = WaveletPhi[HaarWavelet[], x - 1];

w = WaveletPhi[HaarWavelet[], x - 2];

Plot[u, {x, 0, 4}, Exclusions → None];

Plot[v, {x, 0, 4}, Exclusions → None];

Plot[w, {x, 0, 4}, Exclusions → None];

u1 = WaveletPsi[HaarWavelet[], x];

Plot[u1, {x, 0, 3.1}, Exclusions → None];

v1 = WaveletPsi[HaarWavelet[], x - 1];

Plot[v1, {x, 0, 3.1}, Exclusions → None];

w1 = WaveletPsi[HaarWavelet[], x - 2];

Plot[w1, {x, 0, 3.1}, Exclusions → None];

w = WaveletPhi[HaarWavelet[], 2 x];

Plot[w, {x, -1, 2}, Exclusions → None];

w1 = WaveletPsi[HaarWavelet[], 2 x];

Plot[w1, {x, -1, 2}, Exclusions → None];
```

MATLAB Program for Figure 2.4:

```
clc
clear all
  [phi,psi,xval] = wavefun('db3',1);
  plot(xval,phi,'o','MarkerSize',10,...
  'MarkerEdgeColor','red',...
  'MarkerFaceColor',[1 .6 .6])
  xlabel(' x') % x-axis label
  ylabel('Phi') % y-axis label
```

MATLAB Program for Figure 2.5:

```
u = -x^2 + 10 x;

phiD6[x_] = WaveletPhi[DaubechiesWavelet[3], x];

c = {};

For[i = -5; c1, i < 11, i++,
   c1 = NIntegrate[u * phiD6[x - i], {x, -5, 10}, Method → "GaussKronrodRule",
      AccuracyGoal → 6, MaxRecursion → 20, WorkingPrecision → 10]; AppendTo[c, c1];];

u1 = c[[1]] * phiD6[x + 5] + c[[2]] * phiD6[x + 4] + c[[3]] * phiD6[x + 3] +
   c[[4]] * phiD6[x + 2] + c[[5]] * phiD6[x + 1] + c[[6]] * phiD6[x] + c[[7]] * phiD6[x - 1] +
   c[[8]] * phiD6[x - 2] + c[[9]] * phiD6[x - 3] + c[[10]] * phiD6[x - 4] +
   c[[11]] * phiD6[x - 5] + c[[12]] * phiD6[x - 6] + c[[13]] * phiD6[x - 7] +
   c[[14]] * phiD6[x - 8] + c[[15]] * phiD6[x - 9] + c[[16]] * phiD6[x - 10];

Plot[{u, u1}, {x, 0, 10}, Exclusions → None]


phiD4[x_] = WaveletPhi[DaubechiesWavelet[2], x];

d = {};

For[i = -5; d1, i < 11, i++,
   d1 = NIntegrate[u * phiD4[x - i], {x, -5, 10}, Method → "GaussKronrodRule",
      AccuracyGoal → 6, MaxRecursion → 20, WorkingPrecision → 10]; AppendTo[d, d1];];

u2 = d[[1]] * phiD4[x + 5] + d[[2]] * phiD4[x + 4] + d[[3]] * phiD4[x + 3] +
   d[[4]] * phiD4[x + 2] + d[[5]] * phiD4[x + 1] + d[[6]] * phiD4[x] + d[[7]] * phiD4[x - 1] +
   d[[8]] * phiD4[x - 2] + d[[9]] * phiD4[x - 3] + d[[10]] * phiD4[x - 4] +
   d[[11]] * phiD4[x - 5] + d[[12]] * phiD4[x - 6] + d[[13]] * phiD4[x - 7] +
   d[[14]] * phiD4[x - 8] + d[[15]] * phiD4[x - 9] + d[[16]] * phiD4[x - 10];

Plot[{u, u2}, {x, -0.0, 10.0}, Exclusions → None]
```

MATLAB Program for Figure 2.6:

```
In[1]:= phiD6[x_] = WaveletPhi[DaubechiesWavelet[3], x];

In[2]:= Plot[{phiD6[x + 4], phiD6[x + 3], phiD6[x + 2], phiD6[x + 1],
       phiD6[x], phiD6[x - 1], phiD6[x - 2], phiD6[x - 3], phiD6[x - 4]}, {x, -5, 5},
       PlotStyle → {Dashing[Tiny], Dashing[Large], Thickness[0.004], Thickness[0.006],
         Thickness[0.01], Thickness[0.006], Thickness[0.004], Dashing[Large], Dashing[Tiny]},
       PlotLegends → "Expressions", Exclusions → None, PlotRange → All];
```

```
% Example 1 - Boundary Value Problem
clc
close all
clear all

ua=2; % boundary conditions at x=a=0.2
ub=3; % boundary conditions at x=b=0.7

dx=1/63;
x=0:dx:1;
j=6;
m=2^j;

nx=length(x);
[T]=ccm(m,m); %for connection coefficients matrix

f=sin(x*2*pi)';

A=-sin(2*pi*x)/(4*pi^2)+1.904*x+1.6432; %Exact solution of the
  problem.

tic
Tomega=T(:,1);
TomegaCap=fft(Tomega);
fcap=fft(f);
vcap=fcap./TomegaCap;
v=ifft(vcap)/2^(2*j);
toc

%%%%%%%%%%%%%%%%%%%%%%%%%%%%%%%%%%%%%%%%%%%%%%%%%%%%%%%%%%%%%%%%%%%%%%
% v is calculated using fft/ifft which is more efficient than
  the direct
% inverse calculated in the following step.
tic
U=inv(T)*f/2^(2*j); %U and v will be same
toc
%%%%%%%%%%%%%%%%%%%%%%%%%%%%%%%%%%%%%%%%%%%%%%%%%%%%%%%%%%%%%%%%%%%%%%
% Solution for Green function
  f=zeros(nx,1);
  f(1)=1; % delta function
  fcap=fft(f);
  Gcap=fcap./TomegaCap;
  G=ifft(Gcap)/2^(2*j);
  GM=[G(1)  G(33)
      G(33)  G(1)]; % G(a-b)=G(0.7/dt-0.2/dt)=G(45-12)
```

```
  w_a_b=[ua-v(13);ub-v(45)];
  X=inv(GM)*w_a_b;
%%%%%%%%%%%%%%%%%%%%%%%%%%%%%%%%%%%%%%%%%%%%%%%%%%%%%%%%%%%%%%%%%%%%%%%%%%%%

f=zeros(nx,1);
f(13)=X(1);
f(45)=X(2);

fcap=fft(f);
wcap=fcap./TomegaCap;
w=ifft(wcap)/2^(2*j);

  uf=v+w;
  plot(x,A,'r',x,uf,'g.')
  legend('Comparison of exact solution and wavelet solution
  between x=0.2 to x=0.7');
%%%%%%%%%%%%%%%%%%%%%%%%%%% END OF PROGRAM %%%%%%%%%%%%%%%%%%%%%%%%%%%%

%% FUNCTION TO GENERATE CONNECTION COEFFICIENT MATRIX OF
   EXAMPLE -1 %%%%%%%%

function [s] =ccm(m,n)

% These values are written in Page number 264 Wavelet
  Analysis by
% H.L.Resnikoff & R.O.Wells Jr, Springer, Indian Reprint 2008
c0=-5.2679;
c(1)=3.3905;
c(2)=-0.8762;
c(3)=0.1143;
c(4)=0.0054;
c(5:m)=0;

s=zeros(m,n);
for i=1:1:m
  for j=1:1:n
    if j-i==0
          s(i,j)=c0;
    elseif j-i<(m-4) && j>i
          s(i,j)=c(j-i);
    elseif j-i >=(m-4) && j>i
          if j+i-2*(j-m/2)>0
              s(i,j)=c(j+i-2*(j-m/2));
          end
    else
```

```
                    if i-j<(n-4)
                        s(i,j)=c(i-j);
                    else
                        if i+j-2*(i-n/2)>0
                        s(i,j)=c(i+j-2*(i-n/2));
                    end
                end
            end
        end
end
return


%%%%%%%%%%%%%%%%%%%%% END OF CCM FUNCTION %%%%%%%%%%%%%%%%%%%%%

%Example 2 - Initial Value Problem
clc
clear all
close all

order=2; %Select order=1 for first order equation

N=6;
m=N/2; % m-1 is the order of polynomial see eqn 5.16
j=6;
deltaT=2^-j;

t=0:deltaT:0.9999;
y=t.^2-t;
g=y';
k_0=4;
k_f=2^j;

%Daubechies filter coefficients for N=6
a(1)=0.47046720778540;
a(2)=1.14111691583462;
a(3)=0.65036500052742;
a(4)=-0.19093441556852;
a(5)=-0.12083220831070;
a(6)=0.04981749973178;
%%%%%%%%%%%%%%%%%%%%%%%%%%%%%%%%%%%%%%%%%%%%%%%%%%%%%%%%%%%%%%%%
if (order==2)

y_exact=t.^4/12-t.^3/6+0.2*t+1; %for exact solution

o_mega(1)=0.00535714285714; %omega index is varying -4 to 4
                            respectively
o_mega(2)=0.11428571428571; %(-4, -3, -2, -1, 0, 1, 2, 3, 4)
```

```
o_mega(3)=-0.8761907619052;
o_mega(4)=3.39047619047638;
o_mega(5)=-5.26785714285743;
o_mega(6)=3.39047619047638;
o_mega(7)=-0.8761907619052;
o_mega(8)=0.11428571428571;
o_mega(9)=0.00535714285714;

  h=2^(-2*j);
  c(1)=1.0;
  c(2)=1.0031;
  c(3)=1.0062;
  c(4)=1.0094;


%%%%%%%%%%%%%%%%%%%%%%%%%%%%%%%%%%%%%%%%%%%%%%%%%%%%%%%%%%%%%%%%%%%%%%

elseif (order==1)

  y_exact=(t.^3)/3-(t.^2/2)+5;

o_mega(1)=0.003;   %omega index is varying -4 to 4 respectively
o_mega(2)=0.0146; %(-4, -3, -2, -1, 0, 1, 2, 3, 4)
o_mega(3)=-0.1452;
o_mega(4)=0.7452;
o_mega(5)=0.0;
o_mega(6)=-0.7452;
o_mega(7)=0.1452;
o_mega(8)=-0.0146;
o_mega(9)=-0.003;

  h=2^-j;

  c(1)=5.0;
  c(2)=5.0;
  c(3)=4.9998;
  c(4)=4.9993;

end
%%%%%%%%%%%%%%%%%%%%%%%%%%%%%%%%%%%%%%%%%%%%%%%%%%%%%%%%%%%%%%%%%%%%%%
fact_r=zeros(1,m);
mu_n=zeros(1,m);
asum=0;
asum1=0;
for l=0:m-1
fact_r(l+1)=1.0/(2.0*(2.0^l -1.0)); % part of eqn (5.10b)
```

```
if(l==0)
    mu_n(l+1)=1.0;% eqn (5.10a)
end
if(l>=1)
    for i=0:l-1
        for k=0:N-1
            asum=asum+(a(k+1)*k^(l-i));% part of eqn (5.10b)
        end
asum1=asum1+(nchoosek(l,i)*asum*mu_n(i+1));% MATLAB function
  NCHOOSEK(N,K) returns N!/K!(N-K)!
        asum=0;
    end
    mu_n(l+1)=fact_r(l+1)*asum1; %eqn (5.10b)
    asum1=0;
end
end

mu_kl=zeros(N-3,m);

for k=1:N-3
    for l=0:m-1
        for i=0:l
mu_kl(k,l+1)=mu_kl(k,l+1)+(nchoosek(l,i)*k^(l-i)*mu_n(i+1));%eqn
  (5.10c)
        end
    end
end
    mu_negkl=zeros(N-3,m);% mu_negative_kl

    for k=1:N-3
        for l=0:m-1
            for i=0:l
    mu_negkl(k,l+1)=mu_negkl(k,l+1)+(nchoosek(l,i)*(-k)^(l-i)
  *mu_n(i+1));%eqn (5.10c) for -k
            end
        end
    end

    mu_nkl = flipud(mu_negkl); %-1, -2, -3 to -3, -2, -1

    mukl=[mu_n;mu_kl];

    v_ki= mukl*inv(mu_nkl);%eqn (5.1.13)
    omega_neg=[o_mega(4) o_mega(3) o_mega(2) o_mega(1)];

    theta= omega_neg*v_ki;
```

```
%omega_4=o_mega(9), omega_3=o_mega(8), omega_2=o_mega(7),
%omega_1=o_mega(6),omega_0=o_mega(5)
    rho_not=o_mega(5)+theta(3);
    rho_1=(o_mega(6)+theta(2));
    rho_2=(o_mega(7)+theta(1));
    rho_3=o_mega(8);
    rho_4=o_mega(9);


    for p=k_0+1:k_f
c(p)=(h*g(p)-(rho_1*c(p-1)+rho_2*c(p-2)+rho_3*c(p-3)
   +rho_4*c(p-4)))/rho_not;
    end


        plot(t,c,'r+',t,y_exact,'b-');

%          legend(' Exact Solution and Wavelet Solution for Test
   Problem 2');
%%%%%%%%%%%%%%%%%%%%%%%%% END OF PROGRAM %%%%%%%%%%%%%%%%%%%%%%%%%%%%%
```

REFERENCES

1. A. Latto, H.L. Resikoff and E. Tenenbaum (1992), "The evaluation of connection coefficients of compactly supported wavelets," *Proceedings of French-USA Workshop on Wavelets and Turbulence, Princeton University*, June 1991, Springer-Verlag, New York.
2. K. Amaratunga and J.R. Williams (1993), "Wavelet based Green function approach to 2D PDEs," *Engineering Computations*, Vol. 10, No. 4, pp. 349–367.
3. K. Amaratunga, J.R. Williams, S. Qian and J. Weiss (1994), "Wavelet-Galerkin solution for one dimensional partial differential equations," *International Journal of Numerical Methods in Engineering*, Vol. 37, pp. 2703–2716.
4. J. R. Williams and K. Amaratunga (1994), "Introduction to wavelets in engineering," *International Journal for Numerical Methods in Engineering*, Vol. 37, pp. 2365–2388.
5. K. Amaratunga and J.R. Williams (1995), "Time integration using wavelets," *Proceedings of SPIE, Wavelet Applications for Dual Use*, pp. 894–902, Orlando, FL.
6. K. Amaratunga and J.R. Williams (1997), "Wavelet-Galerkin solution of boundary value problems," *Archives of Computational Methods in Engineering*, Vol. 4, No. 3, pp. 243–285.
7. S. Qian and J.J. Weiss (1993), "Wavelets and the numerical solution of partial differential equations," *Journal of Computational Physics*, Vol. 106, No. 1, pp. 155–175.
8. S. Gopalakrishnan and M. Mitra (2010), *Wavelet Methods for Dynamical Problem*, CRC Press, Boca Raton, FL.

CHAPTER 3

Fundamentals of Vector Spaces

T HE CONCEPTS OF FUNCTIONAL ANALYSIS are generally taught to students of mathematics at the graduate level. Most engineering students do not learn this course in a formal way, and therefore these students have no experience with abstract mathematical reasoning. We want to bridge the gap between mathematics and non-mathematics graduates and to help them to grasp the meaning of abstract concepts. This chapter introduces various types of the spaces assuming no prior knowledge in this field. It will serve as a stand-alone introduction to the Hilbert space.

3.1 INTRODUCTION

We consider functional analysis as a wonderful science where we would like to draw the attention of not only engineers and scientists but every thinker whose domain is other than numbers. The concept of spaces, which is the basis of the functional analysis, is used in various fields of life. For example, in the game of football, a coach assigns the space of every player of his team. Similarly, we find the spaces of the army, students, businessmen, etc. Each space is further divided into subspaces which are also spaces. To understand the concept of spaces, consider the examples of mathematics and philosophy, simultaneously.

We try to use the best logic in philosophy and mathematics to understand complex issues. The other means such as observations and experiments have limitations and are often not possible for many problems. The field of mathematicians is the numbers whereas the human being is the field for philosophers. Both have often similar but very difficult questions. Mathematicians try to find the answers to questions like why some particular equations have no solution, why does a method perform very well for one type of equations but does not work well in other class of problems, or how to get the perfect solution without knowing what is the perfect solution. On the other hand, philosophers seek the answer to questions such as what is the purpose of life, reasons for suffering, and how to make this world more peaceful and pleasant.

Mathematicians are able to get perfect and unique answers which are acceptable by all mathematicians in almost all cases. On the other hand, philosophers have many different opinions which often do not converge. The reason is that mathematicians are able to create perfect spaces where they are able to apply logic perfectly whereas philosophers often fail to do the similar exercise. One can see the wonders of very simple mathematical axioms when one tries to understand the concepts of higher dimensions which are beyond the imaginations of the human mind. Simultaneous discussions on mathematical spaces and philosophical spaces will tickle the readers and encourage them to think and participate actively. Like mathematicians, if philosophers develop proper axioms and spaces using some very simple observations, then they can also get wonderful results. Before entering into the domain of mathematical axioms, we shall discuss some philosophy.

In the domain of philosophy, we deal with functions like moral values, happiness of society, suffering of society, individual pleasure, individual distress, self-control, etc., but we do not know properly the independent controlling parameters and the space to maximize or minimize or even find the optimal value. We can observe serious confusion when the method to increase individual pleasure conflicts with the method to increase the happiness of society. The confusion and conflicts of logic are due to lack of well defined spaces.

We find some good examples of axioms, spaces and logic in the Bhagavad Gita, which is one of the oldest texts on philosophy. It starts with the word *dharamshetra*, which means the space of prescribed duty (for individuals). According to this philosophy, the perfect execution of the prescribed duty will lead to Absolute Bliss. Interestingly, the space in Bhagavad Gita is also formed by the concept of axioms.

To understand the formation of axioms, let us recall a common behavior of a society. Disrespecting and considering each other as a fool among religions/societies are very common. As a philosopher, can we define a fool? Our definition of a **fool is: "one who harms oneself."** According to this definition, a drug addict is a fool. We are also fools to some extent because we are continuously harming our environment and so ourselves. But can we call a soldier who is fighting for his country a fool? *A perfect definition or axiom on "self" will help in understanding and defining good and bad for ourselves.* We hope that some axioms will help philosophers to make society more beautiful and predictable. Let us hope that philosophers will achieve perfections like mathematicians. Bhagavad Gita may be a useful reference.

Here, we want to emphasize that the concepts of spaces are not some abstract concepts with little relevance. These concepts exist knowingly or unknowingly in the oldest philosophies in some forms. The greatest form of knowledge is expressed in the form of axioms. It is very exciting to see how mathematical concepts are developed on perfect logic following simple axioms, which leads to the understanding of very complex governing equations. These axioms are independent of dimensions and therefore valid for three-dimensional space as well as infinite dimensional spaces. We start our study on axioms and spaces with the vector spaces [1–5].

3.2 VECTOR SPACES

We study in physics at our school level that a vector is a quantity which has both magnitude and direction. As an example we use displacement as a vector where we use the principles of vector addition to find the resultant of two or more displacements. Do we use angular rotation as a vector because it also has both magnitude and direction? Mathematicians have generalized the familiar concept of the ordinary physical vector to study more general objects and remove shortcomings. The generalization with the help of axioms forms the subject of linear vector space, or vector spaces. The following defines the notion of a vector space. The students should read Appendix A if they are not familiar with sets, field and some common notations.

Definition: Let V be a non-empty set with two operations, namely vector addition and scalar multiplication.

Vector addition: The sum of two elements of the set belongs to the set, i.e., if $u, v \in V$ then $u + v \in V$.

Scalar multiplication: For element α belongs to the field of scalars K, i.e., $\alpha \in K$ and to any $u \in V$ a product $\alpha u \in V$. Then V is called a vector space (over the field K) if the following axioms hold for any vector $u, v, w \in V$

1. $u + v = v + u$

2. $(u + v) + w = u + (v + w)$

3. There is a vector called zero vector (null vector) in V, denoted by 0, such that, for any $u \in V$

$$u + 0 = 0 + u = u$$

4. For each $u \in V$, there is a vector called the negative of u in V, denoted by $-u$, such that

$$u + (-u) = (-u) + u = 0$$

for any scalars $\alpha, \beta \in K$

5. $\alpha(u + v) = \alpha u + \alpha v$

6. $(\alpha + \beta)u = \alpha u + \beta u$

7. $(\alpha\beta)u = \alpha(\beta u)$

8. $1u = u$ for unit scalar $1 \in K$

The first four axioms are concerned with the additive structure of the space V and the last four axioms are concerned with the scalar multiplication on the vector space V.

Example 3.1:

Let V be a set of all elements of the form $(x_1, x_2, x_3, \ldots, x_n)$ for $\forall x_i \in \mathbb{R}$. Define vector addition and scalar multiplication as follows

$$(x_1, x_2, x_3, \ldots, x_n) + (y_1, y_2, y_3, \ldots, y_n) = (x_1 + y_1, x_2 + y_2, x_3 + y_3, \ldots, x_n + y_n)$$

$\alpha(x_1, x_2, x_3, \ldots, x_n) = (\alpha x_1, \alpha x_2, \alpha x_3, \ldots, \alpha x_n) \; \forall \alpha \in \mathbb{R}$. Then V is a linear vector space over $\mathbb{R}$.

Example 3.2:

Let $M_{m \times n}$ be a set of all $m \times n$ matrices over field K. The usual matrix addition and scalar multiplications for any two matrices $A = [a_{ij}]_{m \times n}$, $B = [b_{ij}]_{m \times n} \in M_{m \times n}$ can be expressed as

$$A + B = [a_{ij} + b_{ij}]_{m \times n} \text{ and } \alpha A = [\alpha a_{ij}]_{m \times n}$$

The new matrices are the member of $M_{m \times n}$. It can be easily verified that they satisfy all the axioms of vector space. Therefore, $M_{m \times n}$ is a vector space over field K.

Example 3.3:

Function space is a vector space. For a non-empty set X and any arbitrary field K, $F(X)$ denotes the set of all functions of X into K. $F(X)$ is a vector space because it satisfies the conditions of vector addition, scalar multiplications and other axioms.

If two functions f and $g \in F(X)$ then $f + g \in F(X)$ or

$$(f + g)(x) = f(x) + g(x) \; \forall x \in X$$

The scalar $k \in K$ multiplication to a function $f \in F(X)$ is a function $kf \in F(X)$ or

$$(kf)(x) = kf(x) \qquad \forall x \in X$$

It is not difficult to validate other axioms.

Example 3.4:

If we define vector addition and scalar multiplication for space V with two real numbers $(x_i, y_i) \in V$ as

$$(x_1, \, y_1) + (x_2, \, y_2) = (x_1 x_2, \, y_1 y_2) \text{ and}$$

$\alpha(x, \, y) = (\alpha x^2, \alpha y^2) \; \alpha \in \mathbb{R}$ then V is not a linear vector space.

To verify it, let $(x_1, y_1) \in \mathbb{R}^2, (1,1) \in \mathbb{R}^2$ then
$(x_1, y_1) + (1,1) = (x_1 \cdot 1, \quad y_1 \cdot 1) = (x_1, y_1)$. It is true for all $(x_1, y_1) \in \mathbb{R}^2$.
In this case null vector is $(1,1)$. So axiom (3) is satisfied. Now let us verify axiom (4)

$$(x_1, y_1) + (x_2, y_2) = (x_1 x_2, y_1 y_2)$$

For $x_1 = 0$, we will not get $x_1 x_2 = 1$. Similarly $y_1 = 0$, we will not get $y_1 y_2 = 1$. Therefore, axiom (4) is not satisfied. Hence V is not a vector space.

Example 3.5:

Let V be the set of all solutions of differential equation

$$a\frac{d^2u}{dx^2} + b\frac{du}{dx} + u = 0$$

defined over the interval $[0, 1]$. Then V is a linear vector space because if u_1 and u_2 are solution then $u_1 + u_2$ is the solution and αu_1 is the solution. Other axioms can be tested.

Definition (Subspace): A linear subspace S of a given vector space V over $\mathbb{R}$ is a non-empty subset of V which is itself a linear vector space with respect to the same operation defined over V.

Theorem: A non-empty subset S of V is a subspace of V if and only if for each pair of vector $x, y \in S$ and each scalar $c, d \in K$ such that $cx + dy \in S$.

Example 3.6:

Let $V \in \mathbb{R}^3$, and consider the set

$S = \{x = (x_1, x_2, x_3) \in \mathbb{R}^3 : x_1^2 + x_2^2 + x_3^2 = r^2, r \neq 0\}$. Then S is not a linear subspace of V.
  To verify it, let $(x_1, x_2, x_3) \in \mathbb{R}^3$ and $\alpha \in \mathbb{R}$ then $\alpha(x_1, x_2, x_3) \in \mathbb{R}^3$ but

$$\left(\alpha x_1\right)^2 + \left(\alpha x_2\right)^2 + \left(\alpha x_3\right)^2 = \alpha^2\left(x_1^2 + x_2^2 + x_3^2\right) = \alpha^2 r^2 \neq r^2 \text{ for all } \alpha \in \mathbb{R}.$$

Hence S is not a subspace.

Example 3.7:

Let $V = C^2[0, 1]$ (the space whose all elements can be differentiated twice between 0 and 1) and consider the set $S = \{p(x) \in V : p(0) = \alpha, p(1) = \beta, \alpha, \beta \in \mathbb{R}\}$. Then S is not a linear subspace of V for any nonzero α or β and it is a subspace for $\alpha = \beta = 0$.

  To verify it, let us consider $p(x) \in S$ and $a \in \mathbb{R}$ then $a.p(x) \in C^2[0,1]$, but it is not satisfying the boundary conditions because

$$a.p(0) = a.\alpha \text{ and } a.p(1) = a.\beta$$

so $a \cdot p(x) \notin S$, i.e., it is not satisfying the condition of scalar multiplications.
  Hence S is not linear subspace of V. It can be proved that S is a subspace if $\alpha = \beta = 0$.

Example 3.8:

Let $V = \mathbb{R}^n$ and consider any homogeneous system of linear equations in n unknowns with real coefficients

$$\sum_{j=1}^{n} a_{ij} x_j = 0 \quad (i = 1, 2, 3, \ldots\ldots\ldots, n)$$

for $a_{ij} \in \mathbb{R}$, and $x = (x_1, ..., x_n) \in \mathbb{R}^n$. The set S of all solutions of the homogeneous system is a subspace of V, and is called the *solution space*.

Definition (Linear combination of vectors): Let V be a linear vector space. If an element u in V can be expressed as

$$u = \sum \alpha_i u_i, \alpha_i \in \mathbb{R} \text{ and } u_i \in V,$$

then we say that u is a linear combination of u_i s.

Example 3.9:

Let $u_1 = (1,1,1)$, $u_2 = (1,2,3)$, $u_3 = (5,3,1)$ and $u = \alpha_1 u_1 + \alpha_2 u_2 + \alpha_3 u_3$ for $\alpha_i \in \mathbb{R}$
Here u is a linear combination of u_1, u_2 and u_3.

Example 3.10:

Let $P[x]$ be the vector space of polynomials in x, and let $p_1(x) = x^2 + x + 1$, $p_2(x) = x^2 + x - 1$, $\quad p_3(x) = x^2 - x - 1$, $\quad p_4(x) = x + 1$, $\quad$ and $\quad$ define $p = \alpha_1 p_1 + \alpha_2 p_2 + \alpha_3 p_3 + \alpha_4 p_4$ for $\alpha_i \in \mathbb{R}$, then p is called a linear combination of p_1, p_2, p_3 and p_4.

Definition (Linear independence and dependence of vectors): Let V be a linear vector space. A set of vectors $\{u_1, u_2, ...u_n\}$ are called linearly independent if

$$\alpha_1 u_1 + \alpha_2 u_2 + \alpha_3 u_3 + ...\alpha_n u_n = 0 \text{ iff } \alpha_i = 0, \quad 1 \le i \le n$$

Example 3.11:

$u_1 = (1,1,1)$, $u_2 = (1,2,3)$, $u_3 = (5,3,1)$ are linearly independent

$$\alpha_1(1,1,1) + \alpha_2(1,2,3) + \alpha_3(5,3,1) = 0 \Rightarrow \alpha_1 = \alpha_2 = \alpha_3 = 0.$$

Definition (Span): A set of vectors $\{u_1, u_2, ...u_n\}$ in a vector space V is called a spanning set for V iff every vector $u \in V$ can be written as a linear combination of $u_1, u_2, ...u_n$. We denote the so spanned set simply by $\text{span}\{u_1, u_2, ...u_n\}$. A vector space can have more than one span.

Example 3.12:

Consider the vector space $M_{2\times2}$ of all 2×2 matrices, and the following four matrices

$$A_1 = \begin{bmatrix} 1 & 0 \\ 0 & 0 \end{bmatrix}, A_2 = \begin{bmatrix} 0 & 1 \\ 0 & 0 \end{bmatrix}, A_3 = \begin{bmatrix} 0 & 0 \\ 1 & 0 \end{bmatrix}, A_4 = \begin{bmatrix} 0 & 0 \\ 0 & 1 \end{bmatrix}$$

These four matrices are spanning set for $M_{2\times2}$ because any matrix $A \in M_{2\times2}$ can be written as a linear combination of these four matrices.

Example 3.13:

A vector $u = (\alpha, \beta, \gamma) \in \mathbb{R}^3$ can be written as a linear combination of vectors $\{u_1, u_2, u_4\}$, $\{u_2, u_3, u_4\}$ or $\{u_1, u_2, u_3, u_5\}$, where $u_1 = (1,1,1)$, $u_2 = (0,1,1)$, $u_3 = (1,0,1)$, $u_4 = (1,1,0)$ and $u_5 = (1,2,3)$. We have

$$(\alpha, \beta, \gamma) = (\alpha - \beta + \gamma)u_1 + (\beta - \alpha)u_2 + (\beta - \gamma)u_4$$

$$= \frac{1}{2}\left((-\alpha + \beta + \gamma)u_2 + (\alpha - \beta + \gamma)u_3 + (\alpha + \beta - \gamma)u_4 \right)$$

$$= (\alpha + \beta - \gamma)u_1 + (\gamma - 3\alpha)u_2 - (-\gamma + \alpha + \beta)u_3 + \alpha u_5$$

Definition (Basis): A set $S = \{u_1, u_2, \ldots\ldots\ldots u_n, \ldots\}$ of vectors is a basis of V iff it has the following properties:

1. S is linearly independent, i.e., every finite subset of S is linearly independent.
2. S spans V, i.e., every vector in V can be written as finite linear combination of elements of S.

Note:

1. Basis of a linear vector space is a maximal linearly independent set.
2. Basis of a linear vector space is a minimal spanning set.
3. A vector space may have more than one basis.

Example 3.14:

- $S = \{A_1, A_2, A_3, A_4\}$ forms a basis for $M_{2\times2}$ in Example 3.12.
- $\{u_1, u_2, u_4\}$ and $\{u_2, u_3, u_4\}$ are the basis for $\mathbb{R}^3$ but $\{u_1, u_2, u_3, u_5\}$ does not form a basis in Example 3.13.
- The following vectors forms the basis for $\mathbb{R}^4$

$$\{(1,1,1,1),(0,1,1,1),(0,0,1,1),(0,0,0,1)\}$$

$$\{(1,0,0,0),(0,1,0,0),(0,0,1,0),(0,0,0,1)\}$$

Example 3.15:

Consider the vector space P_3. All polynomials of degree 3 or less are member of this space. The set $\{1, x, x^2, x^3\}$ is a basis for P_3. The set $\{1, 1+x, x+x^2, x^2+x^3\}$ is also a basis for P_3 because it is linearly independent: $c_1(1) + c_2(1+x) + c_3(x+x^2) + c_4(x^2+x^3) = 0$ $\Rightarrow c_1 = c_2 = c_3 = c_4 = 0$, and it spans P_3, i.e., any arbitrary element $p(x) = a_0 + a_1 x + a_2 x^2 + a_3 x^3 \in P_3$ can be uniquely expressed as a linear combination of element of the set $\{1, 1+x, x+x^2, x^2+x^3\}$.

Definition (Dimension of a vector space): If the number of elements in a basis of a vector space is finite then it is called finite dimensional; otherwise, it is infinite dimensional space. The number of elements in the basis (finite dimensional case) is called the dimension of the vector space and is denoted as dimV.

For example, the dimension of P_3 is 4.

Definition (Direct sum): Let S_1 and S_2 be two subspaces of a vector space V. Then V is said to be direct sum of S_1 and S_2 if

$$V = S_1 + S_2 \text{ and } S_1 \cap S_2 = \{0\}$$

And in this case, we can write

$$V = S_1 \oplus S_2$$

Thus V is said to be the direct sum of S_1 and S_2 if every element $v \in V$ can be uniquely written as

$$v = v_1 + v_2 \text{ where } v_1 \in S_1, v_2 \in S_2$$

Example 3.16:

Let $V = \mathbb{R}^3$, $S_1 = \{(x_1, x_2, 0) \in \mathbb{R}^3 : x_1, x_2 \in \mathbb{R}\}$ and $S_2 = \{(0, 0, x_3) \in \mathbb{R}^3 : x_3 \in \mathbb{R}\}$. Then $V = S_1 \oplus S_2$.

Example 3.17:

Let $V = C[0,1]$ be a set of all continuous functions. Let S_1 and S_2 be the set of all even and odd continuous functions over $[0,1]$, respectively. Then $V = S_1 \oplus S_2$.

3.3 NORMED LINEAR SPACES

We know how to calculate and measure lengths in three-dimensional Euclidean space. Mathematics is not limited to the three-dimensional Euclidean space. The concept of norm generalizes the notion of length or magnitude of a vector.

Let V be a linear vector space over field K ($\mathbb{R}$ or $\mathbb{C}$). Define a mapping $\|\cdot\|: V \to R$ satisfying the following conditions:

1. $\|v\| > 0$, $\quad\quad \forall v \in V$ (positivity)

2. $\|v\| = 0$, iff $v = 0$

3. $\|k\,v\| = |k|\|v\|,$ $\qquad \forall k \in K$

4. $\|u+v\| \le \|u\|+\|v\|,$ $\quad \forall u,v \in V$ (Triangle inequality)

Then $\|.\|$ is called norm on V, and $(V, \|.\|)$ is called a normed linear space.

Example 3.18:

The set of all continuous functions on $[a,b]$, i.e., the vector space $C[a,b]$, is a normed linear space with the norm given as

$$\|f\|_\infty = \sup\{|f(t)| : a \le t \le b\} \text{ and}$$

$$\|f\|_p = \left(\int_a^b |f(t)|^p\right)^{1/p}, \text{ for } 1 \le p < \infty$$

Example 3.19:

All n-tuples of real numbers, i.e., the vector space $\mathbb{R}^n$, is a normed linear space with the norms given by

$$\|x\|_2 = \left(\sum_{i=1}^n |x_i|^2\right)^{1/2},$$

$$\|x\|_\infty = \sup|x_i| \ \forall x = (x_1, x_2, \ldots \ldots x_n) \in \mathbb{R}^n.$$

Note: Every non-zero vector v in V can be multiplied by the reciprocal of its length to obtain the unit vector

$\hat{v} = \frac{1}{\|v\|}v$ then norm of unit vector $\|\hat{v}\| = \frac{\|v\|}{\|v\|} = 1.$

This process is called normalizing v.

3.4 INNER PRODUCT SPACES

The concept of norm provides a way of measuring the "length" of a vector, or the difference between two vectors of a linear vector space. Similarly, the concept of an inner product is a means of measuring the "angle" between or "orientation" of two vectors.

An inner product space on a real vector space V (V is a vector space over $\mathbb{R}$) is a bilinear form on $V \times V$ that associates with each pair of vectors $u, v \in V$, a scalar, denoted by (u, v) or $\langle u, v \rangle$ that satisfies the following axioms

1. $\langle u, u \rangle \ge 0$ (Positivity)

   and $\langle u, u \rangle = 0$ iff $u = 0$

2. $\langle u,v \rangle = \langle v,u \rangle$ (Symmetry)

3. (a) $\langle u_1 + u_2, v \rangle = \langle u_1, v \rangle + \langle u_2, v \rangle$ (Additivity) (b) $\langle \alpha u, v \rangle = \alpha \langle u, v \rangle \forall u, v \in V$ and $\alpha \in \mathbb{R}$
(Homogeneous)

Example 3.20:

Let $u = (a_1, a_2, ...a_n)$, and $v = (b_1, b_2, ...b_n) \in \mathbb{R}^n$. A map $\langle .,. \rangle : \mathbb{R}^n \times \mathbb{R}^n \rightarrow \mathbb{R}$ forms an inner product on $\mathbb{R}^n$ as

$$\langle u,v \rangle = u.v = a_1 b_1 + a_2 b_2 + ... + a_n b_n$$

It is also commonly known as dot product or scalar product. It can be defined in matrix form as

$$\langle u,v \rangle = u^T v.$$

The angle between two vectors $u = 2i + 3j$, and $v = 3i - 2j \in \mathbb{R}^2$ can be calculated as

$$\cos(\theta) = \frac{\langle u,v \rangle}{\|u\|\|v\|} = \frac{[2,3] \cdot [3,-2]^T}{\sqrt{2^2 + 3^2}\sqrt{3^2 + (-2)^2}} = 0$$

Example 3.21:

Let $M_{m \times n}$ be the vector space of all real $m \times n$ matrices. For any two matrices $A = [a_{ij}]_{m \times n}$ and $B = [b_{ij}]_{m \times n} \in M_{m \times n}$, an inner product is defined as

$$\langle A,B \rangle = \sum_{i=1}^{m} \sum_{j=1}^{n} a_{ij} b_{ij},$$

and

$$\|A\|^2 = \langle A,A \rangle = \sum_{i=1}^{m} \sum_{j=1}^{n} a_{ij}^2$$

For $A = \begin{bmatrix} 1 & 2 \\ 3 & 4 \end{bmatrix}$ and $B = \begin{bmatrix} 5 & 6 \\ 7 & 8 \end{bmatrix}$ we get $\langle A,B \rangle = 70$ and $\|A\| = \sqrt{30}$.

Example 3.22:

Let $C[a,b]$ be the vector space of all continuous functions on closed interval $[a,b]$. If two functions $f(x), g(x) \in C[a,b]$, then an inner product $\langle .,. \rangle$ on $C[a,b]$ is defined as

$$\langle f,g \rangle = \int_a^b f(x)g(x)dx.$$

Definition (Orthogonality): Let V be an inner product space. The vectors $u, v \in V$ are said to be orthogonal or u is said to be orthogonal to v if

$$\langle u, v \rangle = 0$$

Exercise: Check the orthogonality for the following cases:

1. Let $u = (1, 2, 3)$, $v = (5, 2, -3)$, $w = (2, -2, 2)$ in $\mathbb{R}^3$.
2. For the functions $\sin t$ and $\cos t$ in the space $C[-\pi, \pi]$ find $\langle \sin t, \cos t \rangle =$
$\int_{-\pi}^{\pi} \sin t . \cos t \, dt$.

Note: Every inner product space is a normed linear space, i.e., inner product, can be associated with a norm by defining $\|u\| = \langle u, u \rangle$. In general, the converse is not true. But a normed linear space is an inner product space if the normed linear space satisfies the law of parallelogram:

$$\|u+v\|^2 + \|u-v\|^2 = 2\left(\|u\|^2 + \|v\|^2\right) \text{ for all } u, v \in V \text{ (Parallelogram Law)}.$$

Theorem (Cauchy-Schwarz inequality): For any two vectors u, v in an inner product space $(V, <>)$

$$\left|\langle u, v \rangle\right| \leq \sqrt{\langle u, u \rangle} \sqrt{\langle v, v \rangle}$$

or

$$\left|\langle u, v \rangle\right| \leq \|u\| \|v\|$$

The theorem can be proved in different ways. The students can assume some numerical values for two vectors $u = (a_1, a_2, \ldots a_n)$ and $v = (b_1, b_2, \ldots b_n) \in \mathbb{R}^n$ and verify the equation:

$$\langle u, v \rangle = \sqrt{\sum_{i=1}^{n} a_i b_i} \leq \left(\sum_{i=1}^{n} |a_i|^2\right)^{1/2} \left(\sum_{i=1}^{n} |b_i|^2\right)^{1/2}.$$

We can compare this theorem with $\cos(\theta) = \frac{\langle u, v \rangle}{\|u\| \|v\|}$, as given in Example 3.20. Since $\cos(\theta) \leq 1$ so the Cauchy-Schwarz inequality is also true.

3.5 BANACH SPACES

Banach space is a complete normed space. For the formal definition of complete space, it is necessary to understand the Cauchy sequence.

Definition (Cauchy sequence): A sequence $\{x_n\}$ in a norm space $(X, \|.\|)$ is called a fundamental/Cauchy sequence if

$$\|x_n - x_m\| \to 0 \quad \text{as } n, m \to \infty.$$

Equivalently $\{x_n\}$ is Cauchy in $(X, \|.\|)$ if for every $\varepsilon > 0$ there exists a natural number N such that

$$\|x_n - x_m\| < \varepsilon \text{ for all } n, m \geq N.$$

Definition (Bounded sequence): A sequence $\{x_n\}$ is said to be bounded if there exists a positive real number M such that $|x_n| \leq M$ for all n.

Definition (Convergent sequence): A sequence $\{x_n\}$ in a norm space $(X, \|.\|)$ is said to be convergent if it approaches some limit. Formally, a sequence $\{x_n\}$ converges to the limit x i.e., $\lim_{n \to \infty} x_n = x$ if, for any $\varepsilon > 0$, there exists an natural number N such that $\|x_n - x\| < \varepsilon$ for all $n > N$. If $\{x_n\}$ does not converge, it is said to diverge.

Theorem 3.1: Every convergent sequence is a Cauchy sequence.

Proof: Let $\{x_n\}$ be a convergent sequence in a norm space $(X, \|.\|)$. Suppose $\{x_n\}$ converges to x. So, for $\varepsilon > 0$ there exists a natural number N such that $\|x_n - x\| < \varepsilon / 2$ for all $n > N$. Similarly, for $\varepsilon > 0$ there exists a natural number M such that $\|x_m - x\| < \varepsilon / 2$ for $m > M$. Now, Let $N' = \max(N, M)$ then for all $n, m \geq N'$

$$\|x_n - x_m\| \leq \|x_n - x\| + \|x - x_m\|$$

$$< \frac{\varepsilon}{2} + \frac{\varepsilon}{2} = \varepsilon.$$

Thus $\{x_n\}$ is a Cauchy sequence.

Note: But the converse of the above result need not be true.

Example 3.23:

Let $X = (0, 1)$ be a norm space with usual norm $\|x - y\| = |x - y|$ for all $x, y \in X$. Let $\{x_n\}$ be a sequence defined by $x_n = 1/n$ then $\{x_n\}$ is a Cauchy sequence. But $\{x_n\}$ is not convergent in X because $0 \notin X$.

Example 3.24:

Consider a sequence of real numbers defined recursively as follows:

$$x_1 > 0,$$

$$x_{n+1} = \left(a + x_n\right)^{-1}, \qquad n \geq 1,$$

where $a > 1$ is a fixed real number. Then, $\{x_n\}$ converges.

For $n \geq 2$

$$\left|x_{n+1} - x_n\right| = \left|\frac{1}{a+x_n} - \frac{1}{a+x_{n-1}}\right| = \frac{\left|x_n - x_{n-1}\right|}{(a+x_n)(a+x_{n-1})} \leq \frac{1}{a^2}\left|x_n - x_{n-1}\right|$$

which shows that $\{x_n\}$ converges to some $x \geq 0$.

$$x = \lim x_{n+1} = \frac{1}{a + \lim x_n} = \frac{1}{a+x} \text{ as } n \to \infty$$

$$\text{i.e., } x^2 + ax - 1 = 0.$$

Hence the given sequence converges to $x = (-a + \sqrt{a^2 + 4})/2$. We do not consider the negative value because $x_i > 0$.

Definition (Complete Space): If every Cauchy/fundamental sequence in the norm space $(X, \|\cdot\|)$ converges to an element in X, then X is said to be complete.

Example 3.25:

- The set of rationals Q is not complete because there exists a Cauchy sequence
- 1, 1.4, 1.41, 1.414, 1.4142, 1.41421,....... which converges to $\sqrt{2}$ but $\sqrt{2} \notin Q$.
- The set of real number $\mathbb{R}$ is complete because every Cauchy sequence is convergent.

Definition (Banach space): A normed linear space which is complete with respect to the norm is called Banach space. In other words a Banach space is a normed space in which every Cauchy sequence converges.

The most familiar example of a Banach space is $\mathbb{R}^n$, Euclidian n-dimensional real space.

3.6 HILBERT SPACE

As a human beings, we can see only the three-dimensional Euclidean space $\mathbb{R}^3$, so all our initial concepts are developed in this space. Without any formal training, it may not be difficult to imagine a larger n-dimensional Euclidean space $\mathbb{R}^n$. This exercise may be interesting for philosophers. Hilbert could generalize this concept and put it in the proper mathematical framework. It helped the mathematician to develop theorems which are very useful for many real life problems. A Hilbert space is called separable if the basis functions are countable. Here, mainly separable Hilbert spaces are presented.

Definition (Hilbert space): A Banach space is a complete normed space and a Hilbert space H is a Banach space whose norm is induced by an inner product. In other words, a vector V over the field K is a Hilbert space iff the following two conditions hold:

1. There is an inner product on V

2. Every Cauchy sequence with respect to the induced norm converges in V.

Some common examples of Hilbert spaces are:

1. L^2 **space:** A piecewise continuous f defined on Ω is square integrable if the integral

$$\int_\Omega |f|^2 \, d\Omega < \infty \text{ i.e., finite.}$$

The vector space of all square integrable functions is called L^2 space, and we say $f \in L^2$ [Appendix C]. It is a Hilbert space H where the inner product of the two functions is defined as:

$$(f,g) = \int_\Omega f \, g \, d\Omega \text{ where } \Omega \subset R^n, \ n \geq 1.$$

The L^2 norm of a function is defined by

$$\|f\|_2 = \left(\int_\Omega |f|^2 \, d\Omega \right)^{1/2}$$

2. l^2 **space:** Let V be the vector space of all infinite sequences of real numbers $(a_1, a_2, \ldots)$ satisfying

$$\sum_{i=1}^\infty a_i^2 = a_1^2 + a_2^2 + \ldots < \infty.$$

Such space is called l_2 or l^2 space. Addition and scalar multiplication are defined in V component wise, that is, if

$$u = (a_1, a_2, \ldots) \text{ and } v = (b_1, b_2, \ldots) \text{ then}$$

$$u + v = (a_1 + b_1, a_2 + b_2, \ldots)$$

$$\text{and } ku = (ka_1, ka_2, \ldots.)$$

An inner product in V is defined by

$$\langle u,v \rangle = a_1 b_1 + a_2 b_2 + \ldots = \sum_{n=1}^{\infty} a_n b_n,$$

It can be proved that the above sum converges (finite) absolutely for any pair of points in V. Hence we can say that l_2 is a Hilbert space.

3.7 PROJECTION ON FINITE DIMENSIONAL SPACE

Let $\{\varphi_1,\varphi_2,\varphi_3,\ldots\}$ be the basis function of Hilbert space H, then a function $f \in H$ can be represented as:

$$f = \sum_{i=1}^{n} c_i \varphi_i$$

For the best approximation, we have to select proper c_i.

Theorem 3.2: Let $\{\varphi_1,\varphi_2,\varphi_3,\ldots\}$ be an orthogonal basis of a finite dimensional Hilbert space V. Then the best approximation to vector $f \in V$ is given by

$$\sum_{i=1}^{n} \frac{\langle f,\varphi_i \rangle}{\|\varphi_i\|^2} \varphi_i$$

or

$$f = \frac{\langle f,\varphi_1 \rangle}{\langle \varphi_1,\varphi_1 \rangle}\varphi_1 + \frac{\langle f,\varphi_2 \rangle}{\langle \varphi_2,\varphi_2 \rangle}\varphi_2 + \ldots\ldots\ldots\ldots \frac{\langle f,\varphi_n \rangle}{\langle \varphi_n,\varphi_n \rangle}\varphi_n$$

Example 3.26:

Let us consider a three dimensional vector space $\mathbb{R}^3$. It is a finite dimensional Hilbert space in the usual sense. This consists of the following three vectors in $\mathbb{R}^3$: $u_1 = (1,1,1)$, $u_2 = (-2,1,1)$, and $u_3 = (0,-1,1)$. We can verify that the vectors are orthogonal $\langle u_1,u_2 \rangle = \langle u_3,u_2 \rangle = \langle u_1,u_3 \rangle = 0$ and hence they are linearly independent. Thus the set of three vectors $\{u_1,u_2,u_3\}$ is an orthogonal basis of $\mathbb{R}^3$.

Suppose we want to write $v = (2,5,9)$ as a linear combination of u_1, u_2, u_3.

$$v = x_1 u_1 + x_2 u_2 + x_3 u_3$$

i.e., $(2,5,9) = x_1(1,1,1) + x_2(-2,1,1) + x_3(0,-1,1)$

we can proceed in two ways.

Method 1: We can write it in the following form

$$x_1 - 2x_2 = 2, \ x_1 + x_2 - x_3 = 5, \ x_1 + x_2 + x_3 = 9$$

Using Gauss elimination, we get

$$x_1 = \frac{16}{3}, \ x_2 = \frac{5}{3}, \ x_3 = 2$$

Method 2: Applying Theorem 3.2: $x_i = \dfrac{\langle v, u_i \rangle}{\langle u_i, u_i \rangle}$
We get

$$x_1 = \frac{2+5+9}{1+1+1} = \frac{16}{3}$$

$$x_2 = \frac{-4+5+9}{4+1+1} = \frac{10}{6} = \frac{5}{3}$$

$$x_3 = \frac{-5+9}{1+1} = \frac{4}{2} = 2$$

Example 3.27:

The following two Mathematica programs show the use of non-orthogonal and orthogonal basis functions of Hilbert space for the approximation of a function $f \in L^2$.

A function $f(x) \in L^2$ can be approximated using orthogonal or nonorthogonal basis functions $\{\varphi_1, \varphi_2, \ldots \varphi_n\}$ of the Hilbert space as

$$f(x) \approx \sum_{i=1}^{n} c_i \varphi_i$$

It is necessary to find appropriate c_i for the best approximation. For nonorthogonal basis functions, we can use least error approximation. Define a functional

$$R(c) = \int_{\Omega} \left(f(x) - \sum_{i=1}^{n} c_i \varphi_i \right)^2 dx$$

The best approximation will occur when we minimize $R(c)$ or

$$\frac{\partial R}{\partial c_i} = 0 \qquad \text{for} \quad i = 1, \ldots n$$

Let us consider a function $f(x) = \frac{x}{2-x^2}$ for $-1 \le x \le 1$ and the cubic basis functions $\{1, x, x^2, x^3\}$. Then the functional

$$R(c) = \int_{-1}^{1} \left(f(x) - \left(c_0 + c_1 x + c_2 x^2 + c_3 x^3 \right) \right)^2 dx$$

$\frac{\partial R}{\partial c_0} = 0$ yields

$$c_0 \int_{-1}^{1} dx + c_1 \int_{-1}^{1} x\,dx + c_2 \int_{-1}^{1} x^2\,dx + c_3 \int_{-1}^{1} x^3\,dx = \int_{-1}^{1} f(x)\,dx$$

$$2c_0 + 0 + \frac{2}{3}c_2 + 0 = \int_{-1}^{1} f(x)\,dx$$

Similarly, $\frac{\partial R}{\partial c_i} = 0$ for $i = 1, 2, 3$ will yield

$$0 + \frac{2}{3}c_1 + 0 + \frac{2}{5}c_3 = \int_{-1}^{1} x f(x)\,dx$$

$$\frac{2}{3}c_0 + 0 + \frac{2}{5}c_2 + 0 = \int_{-1}^{1} x^2 f(x)\,dx$$

$$0 + \frac{2}{5}c_1 + 0 + \frac{2}{7}c_3 = \int_{-1}^{1} x^3 f(x)\,dx$$

The solution of the four equations will give c_i for $i = 1, \ldots 4$ for the best approximation.

For the orthogonal basis of Hilbert space, such as Fourier, Legendre, Chebyshev, etc., the function $f(x) \in L^2$ can be approximated as

$$f = \sum_{i=1}^{n} \frac{\langle f, \varphi_i \rangle}{\langle \varphi_i, \varphi_i \rangle} \varphi_i$$

In the case of a Fourier series

$$f(x) \approx \frac{c_0}{2} + \sum_{i=1}^{n} s_i \sin(i\pi x) + c_i \cos(i\pi x)$$

The trigonometric functions are orthonormal on $[-1, 1]$; therefore,

$$c_0 = \int_{-1}^{1} f(x)\,dx$$

$$c_i = \int_{-1}^{1} f(x)\cos(i\pi x)dx$$

$$s_i = \int_{-1}^{1} f(x)\sin(i\pi x)dx$$

The students should use the following Mathematica program for the generation of a function using nonorthogonal and orthogonal basis functions. They can see the convergence of the series as the number of basis functions increased for the approximation of a function. Various types of convergence are presented in Appendix C.

```
u = c0 + c1 * x + c2 * x^2 + c2 * x^3;

ff[x_] = x / (2 - x^2);

Plot[ff[x], {x, -1, 1}];

a = ∫₋₁¹ (x^3) * ff[x] dx;

b = ∫₋₁¹ (x^2) * ff[x] dx;

c = ∫₋₁¹ x * ff[x] dx;

d = ∫₋₁¹ ff[x] dx;

cc = LinearSolve[{{(2/7), 0, (2/5), 0}, {0, (2/5), 0, (2/3)},
        {(2/5), 0, (2/3), 0}, {0, (2/3), 0, 2}}, {a, b, c, d}];

fapprox = cc[[1]] * x^3 + cc[[2]] * x^2 + cc[[3]] * x + cc[[4]];

Plot[{ff[x], fapprox}, {x, -1, 1}]

c0 = ∫₋₁¹ ff[x] dx

c1 = ∫₋₁¹ ff[x] * Cos[π x] dx;

c2 = ∫₋₁¹ ff[x] * Cos[2 π x] dx;

s1 = ∫₋₁¹ ff[x] * Sin[π x] dx;

s2 = ∫₋₁¹ ff[x] * Sin[2 π x] dx;

(* In this case coefficients of cos terms
   are zero therefore not added in the summation*)

fFourier2 = s1 * Sin[π x] + s2 * Sin[2 π x];

Plot[{ff[x], fFourier2}, {x, -1, 1}]
```

```
s3 = ∫₋₁¹ ff[x] * Sin[3 π x] dx;

fFourier2 = s1 * Sin[π x] + s2 * Sin[2 π x] + s3 * Sin[3 π x] ;

Plot[{ff[x], fFourier2}, {x, -1, 1}]

s4 = ∫₋₁¹ ff[x] * Sin[4 π x] dx;

fFourier4 = s1 * Sin[π x] + s2 * Sin[2 π x] + s3 * Sin[3 π x] + s4 * Sin[4 π x] ;


Plot[{ff[x], fFourier4}, {x, -1, 1}]

(* In this program, see the convergence of Fourier series *)
```

3.8 CHANGE OF BASIS—GRAM SCHMIDT ORTHOGONALIZATION PROCESS

The basis vector $\{\varphi_1,\varphi_2,\varphi_3,...\varphi_n\}$ of the Hilbert space can be converted to an orthogonal basis vector with the help of the Gram Schmidt process. The procedure to construct an orthogonal basis $\{\psi_1,\psi_2,\psi_3,...\psi_n\}$ of V can be illustrated as follows:

$$\psi_1 = \varphi_1$$

$$\psi_2 = \varphi_2 - \frac{\langle\varphi_2,\psi_1\rangle}{\langle\psi_1,\psi_1\rangle}\psi_1$$

$$\psi_3 = \varphi_3 - \frac{\langle\varphi_3,\psi_1\rangle}{\langle\psi_1,\psi_1\rangle}\psi_1 - \frac{\langle\varphi_3,\psi_2\rangle}{\langle\psi_2,\psi_2\rangle}\psi_2$$

$$\psi_4 = \varphi_4 - \frac{\langle\varphi_4,\psi_1\rangle}{\langle\psi_1,\psi_1\rangle}\psi_1 - \frac{\langle\varphi_4,\psi_2\rangle}{\langle\psi_2,\psi_2\rangle}\psi_2 - \frac{\langle\varphi_4,\psi_3\rangle}{\langle\psi_3,\psi_3\rangle}\psi_3$$

$$\vdots$$

$$\psi_n = \varphi_n - \frac{\langle\varphi_n,\psi_1\rangle}{\langle\psi_1,\psi_1\rangle}\psi_1 - \frac{\langle\varphi_n,\psi_2\rangle}{\langle\psi_2,\psi_2\rangle}\psi_2 ....- \frac{\langle\varphi_n,\psi_{n-1}\rangle}{\langle\psi_{n-1},\psi_{n-1}\rangle}\psi_{n-1}$$

Example 3.28:

Let V be the vector space of polynomial $f(x)$ with inner product $(f,g) = \int_{-1}^{1} f(x)g(x)dx$. Apply the Gram-Schmidt orthogonalization process to find the orthogonal basis for the subspace S of V spanned by vectors $v_1 = 1, v_2 = x, v_3 = x^2$ and $v_4 = x^3$.

Let $w_1 = v_1 = 1$. Then

$$w_2 = x - \frac{(x,1)}{(1,1)} \cdot 1 = x - \frac{\int_{-1}^{1} x.1 dx}{\int_{-1}^{1} 1.1 dx} \cdot 1 = x - \frac{0}{2} \cdot 1 = x$$

$$w_3 = x^2 - \frac{(x^2,1)}{(1,1)} \cdot 1 - \frac{(x^2,x)}{(x,x)} \cdot x = x^2 - \frac{\frac{2}{3}}{2} \cdot 1 - \frac{0}{\frac{2}{3}} \cdot x = x^2 - \frac{1}{3}$$

$$w_3 = x^3 - \frac{(x^3,1)}{(1,1)} \cdot 1 - \frac{(x^3,x)}{(x,x)} \cdot x - \frac{\left(x^3, x^2 - \frac{1}{3}\right)}{\left(x^2 - \frac{1}{3}, x^2 - \frac{1}{3}\right)} \cdot \left(x^2 - \frac{1}{3}\right)$$

$$= x^3 - 0.1 - \frac{\frac{2}{5}}{\frac{2}{3}} x - 0 \cdot \left(x^2 - \frac{1}{3}\right) = x^3 - \frac{3}{5} x$$

This basis is known as the Legendre basis.

3.9 RIESZ BASES AND FRAME CONDITIONS

To project a function on space like L^2 space using some basis $\{\phi_i(x)\}$ as

$$f = \sum_{i=1}^{\infty} c_i \phi_i,$$

it is necessary to prove that $\{\phi_i(x)\}$ is a basis function. The existence of basis functions can be verified using the conditions known as Riesz basis conditions. A collection of functions $\{\phi_i(x)\}$ on an interval I. is a Riesz basis if

1. $\{\phi_i(x)\}$ is linearly independent
2. There exist constants $0 < A \le B < \infty$ such that for all functions $f(x)$

$$A\|f\|_2^2 \le \sum_{n=1}^{\infty} |\langle f, \phi_i \rangle|^2 \le B\|f\|_2^2$$

Riesz bases are valuable since they can represent any arbitrary functions uniquely but do not require orthogonality. Hence greater flexibility can be exercised in the construction of such bases. An orthogonal basis is also a Riesz basis. In the case of an orthogonal basis $A = B = 1$.

For any $\{\phi_i(x)\}$, not necessarily linearly independent, which satisfy the condition 2 is called a frame. If $\{\phi_i(x)\}$ is a frame, then any function $f(x)$, has representation

$$f(x) = \sum_{i=1}^{\infty} c_i \phi_i(x)$$

If $\{\phi_i(x)\}$ is not linearly independent, then this representation is not unique, which means that coefficients c_i will not be unique.

REFERENCES

1. J.N. Reddy (1986), *Applied Functional Analysis and Variational Methods in Engineering*, McGraw-Hill, New York.
2. B.D. Reddy (1998), *Introductory Functional Analysis: With Applications to Boundary Value Problems and Finite Elements*, Springer, New York.
3. E. Kreyszig (1978), *Introductory Functional Analysis with Applications*, John Wiley & Sons, New York.
4. M.S. Gockenbach (2010), *Finite-Dimensional Linear Algebra*, CRC Press, Boca Raton, FL.
5. Gilbert Strang (2006), *Linear Algebra and Its Applications*, Thomson Brooks/Cole, Noida, India.

CHAPTER 4

Operators

T HE OBJECTIVE OF THIS CHAPTER is to present differential operator where we want to know about the existence of its solution, spectrum of the operator, etc. For completeness, we have introduced other types of operators also. Operator theory is very well developed and used [1–4]. Mass, energy, momentum, etc. are expressed as the linear operators in quantum mechanics. The space of wave function is the Hilbert space, and its dynamics are governed by the Schrödinger equation. This exhaustive theory is beyond the scope of this book. We can consider our body as a kind of operator which transforms one set of input to another set of output. The body can also be considered as a set of operators where the elements are ears, eyes, nose, etc. Input of each of these operators belongs to some space, and the output may be considered as the positive and negative feelings whose values are as continuous as real numbers. The difference between such real life operators and mathematical operators is perhaps that the latter do not "consume" input and the output is also not inconsistent. Here, a real life operator is not our subject, so when we use the term operator, it means mathematical operator only.

4.1 GENERAL CONCEPT OF FUNCTIONS

We all have studied various types of functions. A function establishes relations between any two sets whereas an operator acts on two vector spaces. In other words, function is the generalized concept of the term operator.

A function is a relation (or correspondence) that is "single valued." In other words, a function transforms each element from one set into only one element in other.

Let X and Y be two sets, and suppose that "f" is a rule that assigns to each element in X exactly one element in Y. Then "f" is called a function defined on X with values in Y.

The set X is called the domain of f, denoted by $D(f)$ and Y is called the co-domain of f. If $y = f(x)$, we say y is the image of x. The range of a function $f : X \rightarrow Y$ (means f maps set X into set Y), denoted as $R(f)$, is the set of all images f. If $R(f) \subset Y$, f is said to map X into Y. If $R(f) = Y$, then f is said to map X onto Y.

If the domain of a function f does not contain two elements with the same image, then f is called a one-to-one function. When a function is one-to-one and onto both then there exists a mapping from the range to the domain, called inverse of the function, and the function is said to be invertible.

If the range of a function contains only one element, the function is called a constant function. The terms mapping, transformation and function are often used synonymously.

Operators are used to express mathematical problems generally in the form of equations or functions to be solved or optimized. At the most primitive level, an operator requires only two spaces. Useful mathematical theories of operators are developed for the sets which have the structure of vector spaces. To examine the solvability of a mathematical problem, the properties of the vector spaces have a vital role.

4.2 OPERATORS

For two sets X and Y, a rule T, generally called an operator or transformation or mapping, assigns to each element in a subset of X a unique element in Y. We write an operator T as

$$T : X \rightarrow Y$$

The expression reads "T maps elements of X to elements of Y." It can be also stated as an element $x \in X$ is mapped to an element $y \in Y$. This transformation can be expressed as

$$Tx = y$$

The domain $D(T)$ of T is the subset of X and image space or range $R(T)$ of T is the set of element in Y.

In other words

$$R(T) = \{ y \in Y : Tx = y \text{ for some } x \in D(T) \}$$

Generally the domain $D(T)$ is the whole set X unless it is stated explicitly.

Example 4.1:

Figure 4.1 shows some function $y = f(x)$ defined on the interval $D(f) = [-a, a]$.

Definition (Bijective operator): If the range of T happens to be all Y, i.e., $R(T) = Y$ then T is called a surjective operator, and we say that T maps X onto Y. Otherwise T maps X into Y.

An operator T is said to be one-to-one or injective if no two distinct elements of X are mapped to the same element of Y, i.e., $x_1 \neq x_2 \Rightarrow T(x_1) \neq T(x_2)$.

If T is surjective and is also injective, then T is said to be bijective.

If $T : X \rightarrow Y$ is bijective then its inverse $T^{-1} : Y \rightarrow X$ can be defined as

$$T^{-1}(y) = \{ x \in X : Tx = y \}.$$

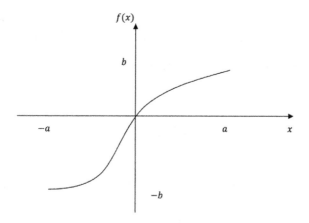

FIGURE 4.1 An operator f with domain $D(f)=[-a,+a]$ and range $R(f)=[-b,+b]$.

Example 4.2:

Let $f : \mathbb{R}^2 \to \mathbb{R}$, $f(u,v) = u^2 + v^2$

Range, i.e., $R(f)$ is only non-negative members; therefore, f is not surjective.

Example 4.3:

The Laplace operator is defined by

$$\nabla^2 u \equiv \frac{\partial^2 u}{\partial x^2} + \frac{\partial^2 u}{\partial y^2} = f$$

For $u \in C^2(\Omega)$ with Ω a domain in $\mathbb{R}^2$, the image f of u is a continuous function. To be specific $\nabla^2 : C^2(\Omega) \to C(\Omega)$. There are infinite numbers of linearly independent solutions for $f = 0$. So the Laplace opertor is not injective and it can be proved for any other f also.

For example if $\Omega \in \mathbb{R}^2$ and $u = x^4 y^3$ or $u = x^4 y^3 + x$ then the image of u is the function f defined by

$$f(x,y) = 12x^2 y^3 + 6x^4 y$$

Definition (Isomorphism): Two vector spaces X and Y are said to be isomorphic to each other if there exists a linear bijective map T from X onto Y. The map is then called an isomorphism. If T is isomorphism then the inverse operator T^{-1} is also isomorphism from Y onto X.

Isomorphism provides information about whether it is possible to put one-to-one correspondence between the elements of two space X and Y. This establishes that the two spaces are "alike" in a rough sense.

Definition (Isometric operator): A bounded operator T on a Hilbert space H is called an isometric operator if $Tx = x$ for all $x \in H$.

Definition (Unitary operator): A bounded operator on a Hilbert space H is called a unitary operator if $T^*T = TT^* = I$ on H, where T^* is the adjoint of T.

Definition (Integral operator): An integral operator $T: C^0[a,b] \to C^1[a,b]$ can be defined as

$$Tu(x) = \int_a^x u(\xi)d\xi \qquad x \in [a,b]$$

Such an integral operator with a variable range of integration is called a Volterra integral operator.

Another type of frequently used integral operators are

$$v = Tu \qquad \text{or} \qquad v(\eta) = \int_a^b k(\eta,\xi)u(\xi)d\xi$$

Here the function $k(\eta,\xi)$ is called the kernel of T. The word kernel is also used for null space which has completely different meaning. There are numerous useful operators such as Fourier and Laplace, specified by a choice of the function k of two variables.

Some kernels have an associated inverse kernel $k^{-1}(\eta,\xi)$ which yields an inverse transform:

$$u(\xi) = \int_a^b k^{-1}(\eta,\xi)v(\eta)d\eta$$

It is called a symmetric kernel if it is unchanged when the two variables are permuted.

Continuous operator: To understand a continuous operator where an element from one norm space is mapped to another norm space, let us recall the definition of a continuous function:

A function $f: \mathbb{R} \to \mathbb{R}$ is said to be continuous at x_0 if for any $\epsilon > 0$, we can find a $\delta > 0$ such that $|f(x_0) - f(x)| < \epsilon$ whenever $|x_0 - x| < \delta$. The function is continuous if it is continuous for all x.

Definition: Let X and Y be two norm spaces. The operator $T: X \to Y$ is said to be continuous at $x_0 \in X$ if for every $\epsilon > 0$ there exists a positive number δ, which may depend on x_0 and ϵ such that

$$\|Tx_0 - Tx\| < \epsilon \quad \text{whenever} \quad \|x_0 - x\| < \delta$$

If it holds for every $x_0 \in X$, then we simply say that T is continuous on X. T is said to be uniformly continuous on X if δ does not depend on x_0.

Definition (Bounded operator): Let X and Y be two normed spaces. The linear operator $T: X \to Y$ is said to be bounded if and only if there exists a constant $\gamma \geq 0$ such that

$$\|Tx\| \leq \gamma \|x\| \quad \forall\, x \in X$$

The norm of T is defined as:

$$\|T\| = Sup\left\{ \frac{\|T\|}{\|x\|}, u \neq 0 \right\}$$

It is not difficult to show that

$$\|Tx\| \leq \|T\| . \|x\|$$

for a bounded linear operator.

"Norm" of T indeed satisfies the norm axioms. The set of all bounded linear operators $\mathcal{L}(X,Y)$ from X to Y forms norm spaces. To verify this, consider T and S, which belong to $\mathcal{L}(X,Y)$, then $\alpha T + \beta S$ is defined by

$$(\alpha T + \beta S)u = \alpha Tu + \beta Su$$

also belongs to $\mathcal{L}(X,Y)$ for all $u \in X$ and $\alpha, \beta \in \mathbb{R}$ or $\mathbb{C}$.

The norm axioms can be verified as:

$$\|T\| \geq 0 \;;\; \|T\| = 0 \text{ if and only if } T = 0 \text{ (zero operator)}$$

The triangular inequality follows from $\|T+S\| = \underset{u\neq 0}{Sup} \dfrac{\|(T+S)u\|}{\|u\|} = \underset{u\neq 0}{Sup} \dfrac{\|(Tu+Su)\|}{\|u\|}$

$$\leq \underset{u\neq 0}{Sup}\left(\frac{\|Tu\|}{\|u\|} + \frac{\|Su\|}{\|u\|} \right) = \|T\| + \|S\|$$

Thus $L(X,Y)$ is a normed space.

Definition (Null space): The null space $N(T)$, also called the kernel of T, Ker (T) is the set of all vectors of $D(T)$ whose image is zero. In other words, the null space is the set of all vectors X such that

$$N(T) = \{x \in X : Tx = 0\}$$

Example 4.4:

The null space of an $m^x n$ matrix A, denoted $Null(A)$, is the set of all solutions to the homogeneous equation $Ax = 0$.

$$Null\, A = \{x : x \in \mathbb{R}^n \text{ and } Ax = 0\}$$

Definition (Projection operators): A linear operator $P : X \to X$, where X is a vector space, is called a projection operator or simply a projection if

$$P(Pu) = Pu \quad \forall u \in X$$

Example 4.5:

Let $u(x)$ be a continuous function in the domain $C[0,l]$. Define the operator $P : C[0,l] \to C[0,l]$ such that

$$Pu = u(0)(l-x)/l + u(l)x/l.$$

P generates the linear interpolation function with two end points. Since

$$P(Pu) = Pu$$

according to the definition of this projection operator, if we generate linear interpolation using the end points of the projection then it will be same as the projection; therefore, P is a projection operator.

Example 4.6:

Consider any vector $x \in \mathbb{R}^3$. The projection operator P projects any vector onto the $\xi\eta$ plane, i.e.,

$$Px = (\xi, \eta, 0)$$

If $x = (\xi_1, \eta_1, \gamma_1)$, then $Px = (\xi_1, \eta_1, 0)$ and $P(Px) = (\xi_1, \eta_1, 0) = Px$. In this case, the range of the operator $R(P) = \{y : y = (\xi, \eta, 0)\}$. For all $y = (0, 0, \gamma)$, $Py = (0,0,0)$, i.e., the null space of P is $N(P) = \{y : y = (0, 0, \gamma)\}$.

Definition (Orthogonal projections): For an inner product space, the projection operator can be expressed as

$$(u, v) = 0 \text{ for } u \in R(P) \text{ and } v \in N(P).$$

The range of the projection operator in inner product space is perpendicular to the null space of P. The situation in $\mathbb{R}^3$ is obvious. Consider the previous example, if P is the projection operator that maps the vector onto the xy-plane, then $R(P)$ is the xy plane, $N(P)$ is the z-axis and $R(P) \perp N(P)$.

4.3 LINEAR AND ADJOINT OPERATORS

Most of problems of engineering are described mathematically by differential, integral or set of algebraic equations. More specifically, consider an operator equation

$$Au = f$$

It can be a differential equation

$$-\frac{d}{dx}\left(a\frac{du}{dx}\right) = f \quad 0 < x < L$$

or a matrix in the finite dimension

$$\begin{bmatrix} a_{11} & a_{12} & . & . & a_{1n} \\ a_{21} & a_{22} & . & . & a_{2n} \\ . & . & . & . & . \\ . & . & . & . & . \\ a_{m1} & a_{m1} & . & . & a_{mn} \end{bmatrix}\begin{bmatrix} u_1 \\ u_2 \\ . \\ . \\ u_n \end{bmatrix} = \begin{Bmatrix} f_1 \\ f_2 \\ . \\ . \\ f_m \end{Bmatrix}$$

Here u belongs to a vector space U ($u \in U$) and f belongs to V. The operator A maps to each element u from a linear vector space U with some element v of linear vector space $V (A : U \rightarrow V)$. The terms transformation, mapping and operator are often used interchangeably.

Definition: The transformation $A : U \rightarrow V$ over the same field F, is said to be linear if

1. $A(\beta u) = \beta A(u) \forall\ u \in U$ and scalar $\beta \in F$
2. $A(u_1 + u_2) = A(u_1) + A(u_2) \forall\ u_1, u_2 \in U$

Theorem 4.1: The linear operator A is one to one if and only if $N(A) = \{0\}$.

Theorem 4.2: A linear operator T from a normed space X to a normed space Y is continuous if and only if it is bounded.

Example 4.7:

Consider the differentiation operator $\frac{d}{dx} : C^1[0,1] \rightarrow C[0,1]$, i.e.,

$$\frac{d}{dx} : f \rightarrow f' \text{ for } f \in C^1[0,1]$$

Domain of the operator, $D\left(\frac{d}{dx}\right) = C^1[0,1]$, which is a proper subspace of $C[0,1]$. It can be verified that the differentiation operator is a surjection, $R\left(\frac{d}{dx}\right) = C[0,1]$. The differentiation operator is not injective because we can find f_1 and f_2 with same derivative f', and its null set is the set of constant functions.

Definition (Adjoint operator): Let U and V be the Hilbert spaces, and let $A:U \to V$ be a linear operator. The adjoint of A is the operator $A^* : V \to U$ that satisfies

$$\left(Au,v\right)=\left(u,A^*v\right) \text{ for all } u \in U, \ v \in V$$

> **Remark 1:** The term "adjoint" for cofactor matrix used for matrix inversion is completely unrelated with this adjoint. The cofactor matrix should be called "classical adjoint" instead of simply "adjoint."
>
> **Remark 2:** If U and V are the normed spaces and $A:U \to V$ is a bounded linear operator, then the adjoint operator associated to A is $A^x : V^* \to U^*$. Here U^* and V^* are dual spaces (defined below) of U and V. The operator A^* is also called a Hilbert-adjoint operator. In the case of Hilbert spaces $U=U^*$.

Example 4.8:

Let A be a real matrix $m \times n$ viewed as a linear operator $A:\mathbb{R}^n \to \mathbb{R}^m$. Then

$$\left(Au,v\right)=\left(Au\right)^T v = u^T A^T v = \left(u,A^T v\right)=\left(u,A^* v\right)$$

Thus the adjoint of a matrix is its transpose. If $A:\mathbb{C}^n \to \mathbb{C}^m$, then the conjugate transpose of A is the adjoint A^*.

Example 4.9:

Let A be a differential operator. The adjoint A^* is the differential operator that satisfies

$$\left(Au,v\right)=(u,A^*v)$$

for sufficiently smooth u and v with compact support in Ω. Let

$$Au(x)=-\frac{d}{dx}\left(a(x)\frac{d}{dx}u(x)\right)+\frac{d}{dx}(b(x)u(x)) \text{ on } [0,1].$$

then

$$\left(Au,v\right)=\int_0^1 \left[-\frac{d}{dx}\left(a(x)\frac{d}{dx}u(x)\right)+\frac{d}{dx}\left(b(x)u(x)\right)\right]v(x)\,dx$$

Integrating by parts gives

$$=\int_0^1 -u(x)\frac{d}{dx}\left(a(x)\frac{d}{dx}v(x)\right)dx+\left(u(x)a(x)\frac{d}{dx}v(x)\right)\Big|_0^1 -\left(v(x)a(x)\frac{d}{dx}u(x)\right)\Big|_0^1$$

$$-\int_0^1 u(x)b(x)\frac{d}{dx}v(x)\,dx+\left(b(x)u(x)v(x)\right)\Big|_0^1 = (u,A^*v)$$

All of the boundary term vanishes due to compact support. Therefore,

$$A^* v = -\frac{d}{dx}\left(a(x)\frac{d}{dx}v(x)\right) - b(x)\frac{d}{dx}(v(x))$$

Definition (Self-adjoint operator): If $T = T^*$, that is, $(Tx, y) = (x, Ty)$ for all $x \in H$, then T is called self-adjoint.

Example 4.10:

The operator T on $L^2(R)$ define by $T(f)(t) = tf(t)$ is self-adjoint. We have $(Tf, g) = \int_R tf(t)\overline{g(t)}dt = \int_R f(t)\overline{tg(t)}dt = (f, Tg)$

Definition (Positive definite operator): An operator T is called positive **definite** if it is self-adjoint and $(Tx, x) \geq 0$, for all $x \in H$.

4.4 FUNCTIONALS AND DUAL SPACE

Linear functional is a special case of linear operator where the operator assigns scalars to vectors. In other words, $V = \mathbb{R}$.

Definition (Linear Functional): Let U be a vector space over the real numbers field $\mathbb{R}$ (or $\mathbb{C}$). A linear transformation l of U into $\mathbb{R}$ (instead of space V) is called a linear form or a linear functional.

$$l : U \to \mathbb{R}$$

Example 4.11:

Let δ_y be the delta function at a point y in a region Ω. It gives a linear functional on sufficiently smooth $u(x)$

$$F(u) = \int_\Omega \delta_{y-x} u(x)dx = u(y)$$

Example 4.12:

Let α in $\mathbb{R}^n$ be fixed. Define a map $l : \mathbb{R}^n \to \mathbb{R}$ given by

$$l(x) = \sum_{i=1}^n \alpha_i x_i, \text{ for all } x = (x_1, x_2, \ldots, x_n) \in \mathbb{R}^n$$

Then l is a linear functional on U.

Example 4.13:

Projection mapping is defined as $\pi_i(\alpha_1, \alpha_2, \ldots \alpha_n) = \alpha_i$, i.e., $\pi_i : \mathbb{R}^n \to \mathbb{R}$. It can be shown that it is a linear functional.

Example 4.14:

Let $I : U \to \mathbb{R}$ be the integral operator and U be the vector space of polynomials over field $\mathbb{R}$. Then integral $\int_\Omega f(x) dx$ is a linear functional.

Theorem 4.3 (Riesz representation): Every bounded linear functional $l(x)$ on a Hilbert space H can be expressed in terms of inner product as

$$l(x) = (y, x) \text{ the unique vector } y \in H.$$

In other words, every bounded linear functional is given by a fixed vector y.

Definition (Dual space): Dual space is a vector space which is a set of all linear functionals. The dual space to a vector space U is defined as the vector space U' or $U^* = \mathcal{L}(U, \mathbb{R}) = \{l : U \to \mathbb{R} | l \text{ is linear}\}$, i.e., dual space consists of all real-valued functionals.

Let U be the finite dimensional space and let $\{x_1, x_2 \ldots x_n\}$ be a basis of U. Then $\{\varphi_1, \varphi_2 \ldots \varphi_n\}$ is the basis of dual space U^* if

$$\varphi_i(x_j) = \delta_{ij} = \begin{cases} 1 & \text{if } i = j \\ 0 & \text{if } i \neq j \end{cases} \tag{4.1}$$

In this case any vector $u \in U$ can be expressed as

$$u = \varphi_1(u) x_1 + \varphi_2(u) x_2 + \ldots + \varphi_n(u) x_n \tag{4.2}$$

and any linear functional $f \in U^*$ can be expressed as

$$f = f(x_1) \varphi_1 + f(x_2) \varphi_2 + \ldots + f(x_n) \varphi_n \tag{4.3}$$

Example 4.15:

Consider a non-orthogonal basis of $\mathbb{R}^2$ $\{x_1 = (1,1), x_2 = (1,2)\}$. Let the dual basis be $\varphi_1 = ax + by$ and $\varphi_2 = cx + dy$. The equation (4.1) leads to the following equations:

$$\varphi_1(1,1) = a + b = 1; \varphi_1(1,2) = a + 2b = 0; \varphi_2(1,1) = c + d = 0; \varphi_2(1,2) = c + 2d = 1;$$

The solutions yield the dual basis $\varphi_1 = 2x - y$ and $\varphi_2 = -x + y$.

Let $u = (4,5)$. Since $u \in \mathbb{R}^2$, it can be expressed as $u = \alpha x_1 + \beta x_2$.

The constants can be obtained by using the duals equation (4.2) as

$$\alpha = \varphi_1(u) = \varphi_1(4,5) = 3$$

and

$$\beta = \varphi_2(u) = \varphi_2(4,5) = 1$$

Now u can be expressed using the primal basis as $u = 3x_1 + x_2$.

Here $\{\varphi_1, \varphi_2\}$ are the basis functions of dual space U^*. Consider an element of dual space $\pi_i(\alpha_1, \alpha_2) \in U^*$. The projection functional $\pi_i(\alpha_1, \alpha_2)$ of $\mathbb{R}^2$ can be expressed as

$$(\pi_1, \pi_2) = c_1 \varphi_1 + c_2 \varphi_2$$

The constants can be obtained by using primal basis functions as equation (4.3)

$$c_1 = \pi_1(x_1) = \pi_1(1,1) = 1$$

$$c_2 = \pi_2(x_2) = \pi_2(1,2) = 2$$

Therefore,

$$(\pi_1, \pi_2) = \varphi_1 + 2\varphi_2 = (x, y)$$

Definition (Bilinear Functional): Let U and V be two vector spaces over the same field $F(\mathbb{R}$ or $\mathbb{C})$ and the operator $a(u, v)$ maps pairs (u, v), $u \in U$, $v \in V$ into the field of scalars F. The operator $a : U \times V \to F$ is called a bilinear functional or bilinear form if $a(u, v)$ satisfies the condition

$$a(\eta u_1 + \beta u_2, \lambda v_1 + \delta v_2) = \eta\lambda \, a(u_1, v_1) + \eta\delta \, a(u_1, v_2) + \beta\lambda \, a(u_2, v_1) + \beta\delta \, a(u_2, v_2)$$

For all $u_1, u_2 \in U$, $v_1, v_2 \in V$ and scalars η, β, λ and δ.

Remark: A bilinear form $a(.,.) : U \times U \to F$ is said to be symmetric if $a(u, v) = a(v, u)$, and skew-symmetric if $a(u, v) = -a(u, v)$ for all $u, v \in U$.

Example 4.16:

Let $\{\varphi_1, \varphi_2, \ldots \ldots \varphi_m\}$ and $\{\psi_1, \psi_2, \ldots \ldots \psi_n\}$ be the basis vectors of finite dimensional spaces U and V, respectively. Any $u \in U$ and $v \in V$ can be expressed as

$$u = \sum_{i=1}^{m} \eta_i \phi_i \text{ and } v = \sum_{j=1}^{n} \beta_j \psi_j \text{ for } \eta_i, \beta_j \in \mathbb{R}.$$

A bilinear form $a(.,.)$ on spaces U and V on finite dimensional spaces is:

$$a(u, v) = a\left(\sum_{i=1}^{m} \eta_i \phi_i, \sum_{j=1}^{n} \beta_j \psi_j\right) = \sum_{i=1}^{m} \sum_{j=1}^{n} \eta_i \beta_j \, a(\phi_i, \psi_j)$$

$$= \sum_{i=1}^{m} \sum_{j=1}^{n} \eta_i \beta_j A_{ij} = \{\eta\}[A]\{\beta\}^T, \quad A_{ij} = a(\phi_i, \psi_j)$$

The matrix $[A]$ is the representation of the bilinear functional $a(u, v)$ in the bases $\{\varphi_i\}$ and $\{\psi_j\}$.

4.5 SPECTRUM OF BOUNDED LINEAR SELF-ADJOINT OPERATOR

The term "spectrum" is motivated by quantum mechanics where spectral energy is characterized as eigenvalue. The spectrum $\sigma(A)$ decomposes the space of a linear operator A into subspaces where the action of A is simple. In the finite dimensional case, the term spectrum refers to the set of eigenvalues of a linear operator. The compact self-adjoint operator is approximated by a finite-dimensional operator; therefore, the spectral theory of both the operators are similar. In general, an operator may have a continuous spectrum besides point spectrum of eigenvalues.

Example 4.17:

Let us consider a square matrix

$$M = \begin{bmatrix} 6 & 0 & 0 \\ 0 & 2 & 5 \\ 3 & 1 & -2 \end{bmatrix}$$

The characteristic equation is

$$\det \begin{bmatrix} 6-\lambda & 0 & 0 \\ 0 & 2-\lambda & 5 \\ 3 & 1 & -2-\lambda \end{bmatrix} = 0$$

$$\Rightarrow \lambda^3 - 6\lambda^2 - 9\lambda - 54 = 0$$

The roots of the polynomial are the eigenvalues: $\lambda_1 = 6, \lambda_2 = -3$ and $\lambda_3 = 3$. In this case, we get three distinct eigenvalues.

Eigenvectors corresponding to eigenvalue $\lambda_1 = 6$ are found by solving the associated homogeneous linear system

$$(M - 6I)x = \begin{bmatrix} 0 & 0 & 0 \\ 0 & -4 & 5 \\ 3 & 1 & -8 \end{bmatrix} \begin{bmatrix} x_1 \\ x_2 \\ x_3 \end{bmatrix} = \begin{bmatrix} 0 \\ 0 \\ 0 \end{bmatrix}, \text{ or } \begin{array}{c} -4x_2 + 5x_3 = 0 \\ 3x_1 + x_2 - 8x_3 = 0 \end{array}$$

We can assume arbitrary real value for, say x_3 and find x_1 and x_2 from the equations. Let $x_3 = 4$, so we get $x_1 = 9$ and $x_2 = 5$.

Thus the eigenvector corresponding to eigenvalue λ_1 is $X = \begin{bmatrix} 9 \\ 5 \\ 4 \end{bmatrix}$.

Similarly, the eigenvector corresponding to eigenvalue $\lambda_2 = -3$ is $Y = \begin{bmatrix} 0 \\ -1 \\ 1 \end{bmatrix}$,

and eigenvector corresponding to eigenvalue $\lambda_3 = 3$ is $Z = \begin{bmatrix} 0 \\ 5 \\ 1 \end{bmatrix}$.

Let us show that these eigenvectors are linearly independent

$$a(9,5,4) + b(0,-1,1) + c(0,5,1) = 0$$

$$\Rightarrow a = 0, \ b = 0 \text{ and } c = 0$$

Hence these vectors are linearly independent. It can be shown that the eigenvectors are the basis vectors of $\mathbb{R}^3$.

Let (x, y, z) be any arbitrary vector in $\mathbb{R}^3$, then it can be represented using these eigenvectors as

$$(x, y, z) = a(9,5,4) + b(0,-1,1) + c(0,5,1)$$

$$\Rightarrow a = \frac{x}{9}, \ b = \frac{1}{18}(-5x - 3y + 15z) \text{ and } c = \frac{1}{6}(-x + y + z)$$

Hence eigenvectors corresponding to different eigenvalues form a basis for the space $\mathbb{R}^3$.

Let us assemble these eigenvectors and form a matrix

$$P = \begin{bmatrix} 9 & 0 & 0 \\ 5 & -1 & 5 \\ 4 & 1 & 1 \end{bmatrix}$$

and the inverse of P is

$$P^{-1} = \begin{bmatrix} \dfrac{1}{9} & 0 & 0 \\ -\dfrac{5}{18} & -\dfrac{1}{6} & \dfrac{5}{6} \\ -\dfrac{1}{6} & \dfrac{1}{6} & \dfrac{1}{6} \end{bmatrix}$$

The matrix M can be diagonalized as

$$P^{-1}MP = \begin{bmatrix} \dfrac{1}{9} & 0 & 0 \\[2mm] -\dfrac{5}{18} & -\dfrac{1}{6} & \dfrac{5}{6} \\[2mm] -\dfrac{1}{6} & \dfrac{1}{6} & \dfrac{1}{6} \end{bmatrix} \begin{bmatrix} 6 & 0 & 0 \\ 0 & 2 & 5 \\ 3 & 1 & -2 \end{bmatrix} \begin{bmatrix} 9 & 0 & 0 \\ 5 & -1 & 5 \\ 4 & 1 & 1 \end{bmatrix} = \begin{bmatrix} 6 & 0 & 0 \\ 0 & -3 & 0 \\ 0 & 0 & 3 \end{bmatrix}$$

The eigenvalues of M are the elements of the diagonal matrix. The diagonalization simplifies many computation processes of the matrix.

It can be proved that if the $n \times n$ matrix M (it may be either real or complex matrix) has n distinct eigenvalues $\lambda_1, \lambda_2, \ldots, \lambda_n$ then the corresponding eigenvectors $v_1, v_2, \ldots, v_n$ form a basis of $\mathbb{R}^n$ or $\mathbb{C}^n$.

If the eigenvector of an $n \times n$ real or complex matrix M span $\mathbb{R}^n$ or $\mathbb{C}^n$, then the matrix is called complete. In particular, any $n \times n$ matrix that has n distinct eigenvalues is complete.

A matrix is real/complex diagonalizable if and only if it is complete.

Example 4.18:

Let us consider a real symmetric matrix

$$M = \begin{bmatrix} 6 & 4 & 8 \\ 4 & 0 & 4 \\ 8 & 4 & -3 \end{bmatrix}$$

The eigenvalues and corresponding eigenvectors for this matrix are

$$\lambda_1 = 13, \ X = \begin{bmatrix} 8 \\ 4 \\ 5 \end{bmatrix},$$

$$\lambda_2 = -8, \ Y = \begin{bmatrix} -2 \\ -1 \\ 4 \end{bmatrix},$$

and

$$\lambda_3 = -2, \ Z = \begin{bmatrix} -1 \\ 2 \\ 0 \end{bmatrix}.$$

It can be verified that these three eigenvectors are orthogonal to each other, i.e.,

$$(8,4,5).(-2,-1,4)=0, (8,4,5).(-1,2,0)=0 \text{ and } (-2,-1,4).(-1,2,0)=0$$

In general, for a real symmetric $n \times n$ matrix, all the eigenvalues of the matrix are real, eigenvectors corresponding to distinct eigenvalues are orthogonal and eigenvectors form the orthogonal basis of $\mathbb{R}^n$.

In particular all symmetric matrices are complete.

A symmetric matrix is called positive definite if all of its eigenvalues are strictly positive.

Example 4.19:

To study the compact self-adjoint operator and its spectral theory, consider a self-adjoint operator A. Let A denote the operator $-\frac{d^2}{dx^2}$, which acts on the function that satisfies the Dirichlet boundary conditions:

$$-\frac{d^2 y}{dx^2} = \lambda y \text{ with boundary conditions } y(0) = y(1) = 0.$$

The differential equation has the form $Ay = \lambda y$. For eigenvalue λ there exists a solution $y \neq 0$.

Let $\lambda = \beta^2$, where $\beta > 0$ then

$$y'' + \beta^2 y = 0$$

This ODE is easy to solve:

$$y(x) = A \cos \beta x + B \sin \beta x$$

where A and B are constants.

Let us impose the boundary conditions,

$$y(0) = 0 = y(1)$$

$$0 = y(0) = A \text{ and } 0 = y(1) = B \sin \beta l$$

We are not interested in $B = 0$, so we must have $\beta l = n\pi$, a root of the sine function. That is

$$\lambda_n = \left(\frac{n\pi}{l}\right)^2, \quad y_n(x) = \sin \frac{n\pi x}{l} \quad (n = 1,2,3,\ldots)$$

In this case $\lambda_n = \left(\frac{n\pi}{l}\right)^2$ are called eigenvalues and the functions $y_n(x) = \sin\left(\frac{xn\pi}{l}\right)$ are called eigenfunctions. For the differential operator that we are interested in, there are an infinite number of eigenvalues $\frac{\pi^2}{l^2}, \frac{4\pi^2}{l^2}, \frac{9\pi^2}{l^2},\ldots$ Thus we can say that we are dealing with infinite dimensional linear algebra.

It can be verified that the eigenfunctions are orthogonal, i.e., $y_i, y_j = \delta_{ij}$, and these eigenfunctions form the basis function of Hilbert space.

4.6 CLASSIFICATION OF DIFFERENTIAL OPERATOR

Consider the general partial differential equations of second order for a function of two independent variables x and y in the form:

$$Rr + Ss + Tt + g(x, y, z, p, q) = 0 \tag{4.4}$$

where

$$p \equiv \frac{\partial}{\partial x}, \quad q \equiv \frac{\partial}{\partial y}, r \equiv \frac{\partial^2}{\partial x^2}, s \equiv \frac{\partial^2}{\partial x \partial y}, t \equiv \frac{\partial^2}{\partial y^2}$$

And R, S and T are continuous functions of x and y.

Then the equation (4.4) is said to be

1. Hyperbolic at a point (x, y) in domain D if $S^2 - 4RT > 0$

2. Parabolic at a point (x, y) in domain D if $S^2 - 4RT = 0$

3. Elliptic at a point (x, y) in domain D if $S^2 - 4RT < 0$

Example 4.20:

Consider the one-dimensional wave equation

$$\frac{\partial^2 z}{\partial x^2} = \frac{\partial^2 z}{\partial y^2} \text{ i.e., } r - t = 0.$$

Now from equation (4.4) we have

$$R = 1, S = 0 \text{ and } T = -1$$

Hence $S^2 - 4RT = 0 - (4 \times 1 \times (-1)) = 4 > 0$. And so the given equation is hyperbolic.

Example 4.21:

Consider the one-dimensional diffusion equation

$$\frac{\partial^2 z}{\partial x^2} = \frac{\partial z}{\partial y} \text{ i.e., } r - q = 0.$$

Now from equation (4.4) we have

$$R = 1 \text{ and } S = T = 0$$

Hence $S^2 - 4RT = 0 - (4 \times 0) = 0$, so the given equation is parabolic.

Example 4.22:

Consider the two-dimensional Laplace's equation

$$\frac{\partial^2 z}{\partial x^2} + \frac{\partial^2 z}{\partial y^2} = 0 \text{ i.e., } r + t = 0.$$

Now from equation (4.4) we have

$$R = 1, S = 0 \text{ and } T = 1$$

Hence $S^2 - 4RT = 0 - (4 \times 1 \times 1) = -4 < 0$, so the given equation is elliptic.

4.7 EXISTENCE, UNIQUENESS, AND REGULARITY OF SOLUTION

Let us consider a differential operator $Au = f$ with the boundary conditions $Bu = g$. To solve this differential equation, we express u in terms of some basis functions (like Chebyshev, Legendre, etc.) or we use difference equation in finite difference method. The process transforms differential equations into a set of algebraic equations.

For example, if φ_i is the basis function, then

$$u = \sum c_i \varphi_i$$

Using the collocation method, it can be expressed at node j as

At node j: $\left(A \sum c_i \varphi_i \right)_j = f_j, \ j = 1, 2, \ldots\ldots\ldots\ldots J$

At boundary points k: $\left(B \sum c_i \varphi_i \right)_k = g_k, \ k = 1, 2, \ldots\ldots\ldots\ldots K$

This can be expressed in matrix form as

$$
\begin{bmatrix}
a_{11} & a_{12} & . & . & . & a_{1n} \\
. & . & . & . & . & . \\
. & . & . & . & . & . \\
. & . & . & . & . & . \\
. & . & . & . & . & . \\
a_{n1} & a_{n2} & . & . & . & a_{nn}
\end{bmatrix}
\begin{bmatrix}
c_1 \\
c_2 \\
. \\
. \\
. \\
c_n
\end{bmatrix}
=
\begin{bmatrix}
y_1 \\
y_2 \\
. \\
. \\
. \\
y_n
\end{bmatrix}
$$

The solution of this finite dimensional matrix operator is the approximate solution of the differential operator. In this section, we are interested to examine the existence and uniqueness of solutions of differential operators. Since the two operators are equivalent, we will start with matrix operators.

Example 4.23:

Let the matrix $M : \mathbb{R}^2 \to \mathbb{R}^2$ be

$$M = \begin{bmatrix} 1 & 2 \\ 9 & 6 \end{bmatrix}$$

In other words, we have the following set of equations

$$\begin{bmatrix} 1 & 2 \\ 9 & 6 \end{bmatrix} \begin{bmatrix} c_1 \\ c_2 \end{bmatrix} = \begin{bmatrix} y_1 \\ y_2 \end{bmatrix}$$

For any real values of y_1 and y_2, we find the unique solution c_1 and c_2. It can be visualized as two lines intersecting at (c_1, c_2). This can also be written as

$$\begin{bmatrix} 1 \\ 9 \end{bmatrix} c_1 + \begin{bmatrix} 2 \\ 6 \end{bmatrix} c_2 = \begin{bmatrix} y_1 \\ y_2 \end{bmatrix}$$

In this case, a column space is created using column vectors, which forms the basis vector of the space. This basis vector can be considered as equivalent to basis function φ_i. Let us change the matrix as

$$M = \begin{bmatrix} 1 & 2 \\ 4 & 8 \end{bmatrix}$$

Geometrically, it can be visualized as two parallel lines. There are two possibilities in this case:

1. The solution does not exist if $y_2 \neq 4 y_1$.
2. The solution is not unique if $y_2 = 4 y_1$. In other words, the lines are coinciding, therefore it has many solutions.

Example 4.24:

We again modify the matrix and consider $M : \mathbb{R}^2 \to \mathbb{R}^3$

$$\begin{bmatrix} a_1 & b_1 \\ a_2 & b_2 \\ a_3 & b_3 \end{bmatrix} \begin{bmatrix} c_1 \\ c_2 \end{bmatrix} = \begin{bmatrix} y_1 \\ y_2 \\ y_3 \end{bmatrix}$$

Can we get solution $[c]$ for any matrix $[M]$ and any vector $[y]$? It can be visualized as the planes formed by two column vectors. It can be expressed as

$$\begin{bmatrix} a_1 \\ a_2 \\ a_3 \end{bmatrix} c_1 + \begin{bmatrix} b_1 \\ b_2 \\ b_3 \end{bmatrix} c_2 = \begin{bmatrix} y_1 \\ y_2 \\ y_3 \end{bmatrix}$$

The equation is

$$\begin{bmatrix} 1 & 2 \\ 9 & 6 \\ 7 & 2 \end{bmatrix} \begin{bmatrix} c_1 \\ c_2 \end{bmatrix} = \begin{bmatrix} y_1 \\ y_2 \\ y_3 \end{bmatrix}$$

If we use the Gauss elimination process, we get

$$\begin{bmatrix} 1 & 2 \\ 9 & 6 \\ 0 & 0 \end{bmatrix} \begin{bmatrix} c_1 \\ c_2 \end{bmatrix} = \begin{bmatrix} y_1 \\ y_2 \\ y_3 - y_2 + 2y_1 \end{bmatrix}$$

In this case the solution exists if it satisfies the relation $y_3 - y_2 + 2y_1 = 0$. This is the essential condition for the existence of a solution.

The equation is again changed as

$$\begin{bmatrix} 1 \\ 9 \\ 7 \end{bmatrix} c_1 + \begin{bmatrix} 2 \\ 6 \\ 2 \end{bmatrix} c_2 + \begin{bmatrix} 3 \\ 15 \\ 9 \end{bmatrix} c_3 = \begin{bmatrix} y_1 \\ y_2 \\ y_3 \end{bmatrix}$$

In this case also the solution exists if $y_3 - y_2 + 2y_1 = 0$, but the solution is not unique. For example, if we assume $y_1 = 4$, $y_2 = 24$ and $y_3 = 16$ then

$$\begin{bmatrix} c_1 \\ c_2 \\ c_3 \end{bmatrix} = \begin{bmatrix} 2 \\ 1 \\ 0 \end{bmatrix} + t \begin{bmatrix} 1 \\ 1 \\ -1 \end{bmatrix}$$

It is true for any value of $t \in \mathbb{R}$. In this case $\begin{bmatrix} 2 & 1 & 0 \end{bmatrix}^T$ is called a particular solution and $t\begin{bmatrix} 1 & 1 & -1 \end{bmatrix}^T$ is known as a complementary solution.

There is a better method to find a complementary solution than what was presented earlier. Consider the solution of a homogeneous equation or null space

$$\begin{bmatrix} 1 & 2 & 3 \\ 9 & 6 & 15 \\ 7 & 2 & 9 \end{bmatrix} \begin{bmatrix} c_1 \\ c_2 \\ c_3 \end{bmatrix} = \begin{bmatrix} 0 \\ 0 \\ 0 \end{bmatrix}$$

Using, Gauss elimination we get

$$\begin{bmatrix} 1 & 2 & 3 \\ 0 & -12 & -12 \\ 0 & -12 & -12 \end{bmatrix} \begin{bmatrix} c_1 \\ c_2 \\ c_3 \end{bmatrix} = \begin{bmatrix} 0 \\ 0 \\ 0 \end{bmatrix}$$

We find the solution $c_1 = c_2 = -c_3$.

So the complementary solution is

$$\begin{bmatrix} c_1 \\ c_2 \\ c_3 \end{bmatrix} = t \begin{bmatrix} 1 \\ 1 \\ -1 \end{bmatrix}$$

Example 4.25:

The Gauss elimination method can be used to find the complementary and particular solutions. Let us consider the following set of equations:

$$\begin{bmatrix} 1 & 2 & 4 & 3 \\ 2 & 4 & 9 & 8 \\ 3 & 6 & 13 & 11 \end{bmatrix} \begin{bmatrix} c_1 \\ c_2 \\ c_3 \\ c_4 \end{bmatrix} = \begin{bmatrix} y_1 \\ y_2 \\ y_3 \end{bmatrix}$$

or

$$\begin{bmatrix} 1 & 2 & 4 & 3 \\ 0 & 0 & 1 & 2 \\ 0 & 0 & 1 & 2 \end{bmatrix} \begin{bmatrix} c_1 \\ c_2 \\ c_3 \\ c_4 \end{bmatrix} = \begin{bmatrix} y_1 \\ y_2 - 2y_1 \\ y_3 - 3y_1 \end{bmatrix}$$

or

$$\begin{bmatrix} 1 & 2 & 4 & 3 \\ 0 & 0 & 1 & 2 \\ 0 & 0 & 0 & 0 \end{bmatrix} \begin{bmatrix} c_1 \\ c_2 \\ c_3 \\ c_4 \end{bmatrix} = \begin{bmatrix} y_1 \\ y_2 - 2y_1 \\ y_3 - y_1 - y_2 \end{bmatrix}$$

In this case, the solution will exist if $y_3 - y_1 - y_2 = 0$. This process is used to find the pivots and free variables. The pivots are the first non-zero entries in their rows and column of zero exists below each pivot. Therefore, c_1 and c_3 are pivot variables and the other variables are called free variables. The free variables are used to determine the pivot variables. To find pivot in terms of free variables, we use one more step of Gauss-elimination:

$$\begin{bmatrix} 1 & 2 & 0 & -5 \\ 0 & 0 & 1 & 2 \\ 0 & 0 & 0 & 0 \end{bmatrix} \begin{bmatrix} c_1 \\ c_2 \\ c_3 \\ c_4 \end{bmatrix} = \begin{bmatrix} 9y_1 - 4y_2 \\ y_2 - 2y_1 \\ y_3 - y_1 - y_2 \end{bmatrix}$$

The pivot variables can be expressed as (obtained from homogeneous equations)

$$c_1 = -2c_2 + 5c_4 \quad \text{and} \quad c_3 = -2c_4$$

Let $c_2 = t$ and $c_4 = s$. If we assume $\begin{bmatrix} y_1 \\ y_2 \\ y_3 \end{bmatrix} = \begin{bmatrix} 6 \\ 9 \\ 15 \end{bmatrix}$;

Then

$$\begin{bmatrix} c_1 \\ c_2 \\ c_3 \\ c_4 \end{bmatrix} = \begin{bmatrix} 18 \\ 0 \\ -13 \\ 0 \end{bmatrix} + t \begin{bmatrix} -2 \\ 1 \\ 0 \\ 0 \end{bmatrix} + s \begin{bmatrix} 5 \\ 0 \\ -2 \\ 1 \end{bmatrix}$$

Here $\begin{bmatrix} 18 & 0 & -13 & 0 \end{bmatrix}^T$ is a particular solution and the free variables t and s give a non-unique complementary solution.

The following theorem gives an alternative method of finding the necessary condition for the existence of the solution:

Theorem 4.4: Let A be a closed bounded below linear operator on Hilbert space. Then $Au = f$ has a solution if and only if $< f, \beta >= 0$, where β is the solution of null space of adjoint operator A^*, i.e., $A^* \beta = 0$.

Example 4.26:

We apply the theorem in the above example. The adjoint matrix $M^* = M^T$ is

$$M^* = \begin{bmatrix} 1 & 2 & 3 \\ 2 & 4 & 6 \\ 4 & 9 & 13 \\ 3 & 8 & 11 \end{bmatrix}$$

Using Gauss elimination method, we get

$$\begin{bmatrix} 1 & 2 & 3 \\ 0 & 0 & 0 \\ 0 & 1 & 1 \\ 0 & 0 & 0 \end{bmatrix} \begin{bmatrix} \beta_1 \\ \beta_2 \\ \beta_3 \end{bmatrix} = \begin{bmatrix} 0 \\ 0 \\ 0 \end{bmatrix}$$

So the free variable is β_3, and pivots are β_1 and β_2. Therefore, the null space solution is

$$\begin{bmatrix} \beta_1 \\ \beta_2 \\ \beta_3 \end{bmatrix} = t \begin{bmatrix} 1 \\ 1 \\ -1 \end{bmatrix}$$

for the existence of solution

$$< y, \beta >= 0$$

Therefore, $y_1\beta_1 + y_2\beta_2 + y_3\beta_3 = 0$

or

$$t(y_1 + y_2 - y_3) = 0$$

or

$$y_1 + y_2 - y_3 = 0.$$

In the case of an operator in the finite dimensional space, the input and output vector belong to finite dimensional real or complex vector spaces. For example, in the case of a set of algebraic equations $M\{c\} = \{y\}$, the size of the matrix is $m \times n$, and the vectors $\{c\} \in \mathbb{R}^n$ and $\{y\} \in \mathbb{R}^m$. But for the differential operator $Au = f$, we have to select continuous functions u and f from the proper spaces. Let A be the differential operator of order $2m$, and if we assume $u \in H^s$, then for continuous dependence on the data f

$$\|u\|_{H^s} \le c\|f\|_{H^{s-2m}}$$

It means that f should not belong to H^n for $n < s - 2m$. The relation states that for a small change in data f, change in u should be small, i.e.,

$$\|\Delta u\|_{H^s} \le c\|\Delta f\|_{H^{s-2m}}$$

The requirement of the space H^s depends on the method which is used to solve the differential operator.

Example 4.27:

This powerful principle can be illustrated with the help of the following example. Let us consider the differential equation,

$$-\frac{d^2u}{dx^2} = \hat{f}, \quad 0 < x < L$$

Due to the second order derivative, the largest possible space to select $u \in H^s$ is $s = 2$ and the other spaces $s > 2$ will be subspaces of H^2. In this case $\hat{f} \in H^0$, i.e., the function must be square integrable. If $\hat{f}$ is a concentrated load, then the problem cannot be solved in this form. For the case of concentrated load, we use the weak form or potential energy form of the equation

$$\int_0^L \left(\frac{du}{dx}\frac{dv}{dx} - \hat{f}v \right) dx - v\frac{du}{dx}\Big|_0^L = 0$$

Now we can select $u \in H^1$ and solve the problem of concentrated load $\hat{f} \in H^{-1}$.

When the differential operator is transformed into a set of linear algebraic equations, the boundary conditions also become a part of the matrix. In this case, we do not study the effect of boundary conditions on the existence and uniqueness of the solution. But in the case of a differential operator, we can see the effect of boundary conditions in the existence and uniqueness of the solution.

We have seen that if β is the solution of finite dimensional homogeneous adjoint operator $([M]^* \{\beta\} = 0)$, then a necessary condition for the solution of $[M]\{c\} = \{y\}$ is that data $\{y\}$ should be orthogonal to $\{\beta\}$, i.e.,

$$\langle y, \beta \rangle = 0.$$

In general, we can say that for checking the existence of the solution for $Au = f$, we have to find the solution of null space of adjoint operator A^*. If the solution of $A^*\beta = 0$ is nonzero then f should be orthogonal to β or $\langle f, \beta \rangle = 0$ for the existence of a solution. In other words, the range of A should be the orthogonal to the null space of adjoint operator $R(A) = N(A^*)^\perp$. Since the null space of adjoint operator $N(A^*)$ will depend on the boundary conditions, $R(A)$ will also depend on the boundary conditions.

Example 4.28:

Consider the differential equation,

$$-\frac{d^2u}{dx^2} = \hat{f} \quad 0 < x < L$$

subject to the boundary conditions,

$$\left(\frac{du}{dx} \right)\Big|_{x=0} = F_1 \quad \left(\frac{du}{dx} \right)\Big|_{x=L} = F_2$$

This can be considered as the axial deformation u of a bar which is subjected to the uniformly distributed load $\hat{f}$, and both the ends are subjected to the forces F_1 and F_2. The equation is self-adjoint, therefore the homogeneous adjoint equation becomes

$$-\frac{d^2 v}{dx^2} = 0$$

with the boundary condition

$$\left(\frac{dv}{dx}\right)\Bigg|_{x=0} = 0 \left(\frac{dv}{dx}\right)\Bigg|_{x=L} = 0$$

Here, $v = \alpha_1 + \alpha_2 x$ is the solution to the homogeneous adjoint equation, where α_1 and α_2 are constants, which we find from the boundary conditions. The two boundary conditions give $\alpha_2 = 0$. According to the theorem, for the existence of a solution,

$$\langle f, v \rangle = 0$$

In this case $f = \hat{f} + \delta(0)F_1 + \delta(L)F_2$

Therefore,

$$\int_0^L \alpha_1 \left(\hat{f} + \delta(0) \, F_1 + \delta(L) \, F_2 \right) dx = 0$$

or

$$\int_0^L \hat{f} \, dx + F_1 + F_2 = 0$$

This relation states that forces must be in equilibrium for the existence of the solution.

REFERENCES

1. J.N. Reddy (1986), *Applied Functional Analysis and Variational Methods in Engineering*, McGraw-Hill, New York.
2. B.D. Reddy (1998), *Introductory Functional Analysis: With Applications to Boundary Value Problems and Finite Elements*, Springer, New York.
3. E. Kreyszig (1978), *Introductory Functional Analysis with Applications*, John Wiley & Sons, New York.
4. J.K. Hunter and B. Nachtergaele (2001), *Applied Analysis*, World Scientific, Singapore.

CHAPTER 5

Theoretical Foundations of the Finite Element Method

I N CHAPTER 1, WE STUDIED THE steps of the finite element method which are used to solve engineering problems. There are many questions related to its theory which need to be answered to avoid mistakes in some of the challenging problems. Some of the issues are discussed in this chapter. For example:

- Is it not incorrect to break a continuous field variable u and use a set of u^e in the potential energy function or functional?

- Can we solve any differential equation using FEM?

- What is the need of converting differential equations in potential energy or functional form for finding a stiffness matrix? Do we have any advantage over other numerical methods such as finite difference and collocation method?

- How do the concepts of functional space help in the solution?

- We have seen that FEM gives the approximate solution and as we increase the number of elements we hope to get improved solution. Is there any method to find error in the solution particularly when the exact solution is unknown? How do we reduce error in an efficient way?

These questions can be addressed appropriately after a brief introduction of Sobolev spaces.

5.1 DISTRIBUTION THEORY

This section presents a short elementary introduction to the theory of distribution particularly for engineers and scientist who have the constraints of basic knowledge and time. Since this book is intended for solution of partial differential equations using FEM and wavelets, a significant part of the theory of distribution is omitted.

Definition (Classical solution): If function u is sufficiently differentiable (in a classical sense) so that

$$\sum_{|\alpha| \leq m} a_\alpha D^\alpha u$$

is well defined and the equation

$$Lu = \sum_{|\alpha| \leq m} a_\alpha D^\alpha u = f \tag{5.1}$$

is satisfied, then u is called a classical solution. (See Appendix C for notations.)

Our task is to find the function u that satisfies a partial differential equation and boundary conditions. The spaces $C^m(\Omega)$ would appear to be appropriate for classical solution of the differential equation which requires m-times continuously differentiable function. However, in most of the real-life problems we find concentrated loads whose classical solution cannot be found, and also it is difficult to satisfy the complex boundary conditions using $u \in C^m(\Omega)$. Therefore, instead of finding a solution in the classical sense, we use the weak formulation of the problem. The weak formulation, which lies at the heart of our finite element method, simplifies the treatment of complex boundary conditions and eliminates the problems of concentrated loads. In order to understand weak formulation, the theory of distributions will be useful.

To show how a problem with highly discontinuous function can be solved with the help of distribution theory, we need to properly define the test function and its space.

In mathematics, often it is advantageous to work with the inner product of function and test function. Let $u(x)$ be a function and $\varnothing$ be a test function (a fuller definition $\varnothing(x)$ will be given later); then $u(x)$ generates distribution U as

$$U(\varnothing) = \langle u, \varnothing \rangle = \int_\Omega u \varnothing \, dx$$

It is generally preferred to write $\langle U, \varnothing \rangle$ instead of $U(\varnothing)$. This is an interpretation of functions as linear functional acting on a space of test functions. This is a very useful concept where dealing with the function $u(x)$ or its differentiation is a tedious job.

Now it raises questions such as what should be the properties of the test function and how the process will help in the case of discontinuous function.

Definition (Test function): The test functions are a special class of infinitely differentiable functions on $\mathbb{R}^n$ with compact support (identically zero outside some closed ball). In other words, the derivatives of all orders of the test functions assume non-zero value only on some compact subset. It is conventional to use the notation $\mathcal{D}(\mathbb{R}^n)$ or simply by $\mathcal{D}$ for the space of all test functions.

Example 5.1: Consider a test function

$$\varnothing(x)=\begin{cases} 0 & |x|\geq a \\ exp\dfrac{1}{x^2-a^2} & |x|<a \end{cases}$$

It can be easily observed that $\varnothing$ is infinitely differentiable and all its derivatives are non-zero in the set $(-a,a)$. The test function on $\mathbb{R}^n$ can be obtained as $\varnothing=\varnothing(x_1)\varnothing(x_2)\dots\varnothing(x_n)$.

To see how the process helps in the case of discontinuous function, let us consider integration-by-parts formula for one-dimension

$$\int_a^b \varnothing u'dx=\left[\varnothing u\right]_a^b-\int_a^b u\varnothing'dx$$

Since $\varnothing=0$ on the boundary, the equation reduces to

$$\int_a^b \varnothing u'dx=-\int_a^b u\varnothing'dx$$

Since $\varnothing'$ exists, we get the solution. This can be easily generalized to a partial derivative of order m for higher dimension. Let

$$Lu=\sum_{|\propto|\leq m}a_\propto D^\propto u=f$$

In many cases, we may not find continuous functions u and f to fit into this equation. For example f may be concentrated force acting on a point and u may not be differentiable in the classical sense up to the order m.

Let $\varnothing$ be a test function on $\Omega\subseteq\mathbb{R}^n$. The equation is multiplied by $\varnothing$ and integrated over the domain

$$\int_\Omega \varnothing\left(\sum_{|\propto|\leq m}a_\propto D^\propto u-f\right)d\Omega=0 \tag{5.2a}$$

Since $\varnothing(x)=0$ outside a bounded subset on Ω, integration by parts yields

$$(-1)^{|\alpha|} \int_{\Omega} \left(\sum_{|\alpha| \leq m} a_\alpha u D^\alpha \emptyset \right) d\Omega = \int_\Omega f \emptyset \, d\Omega \qquad (5.2b)$$

Now the derivatives of u are not required and f is such that distribution (right hand side) exists.

Definition (Convergence of test functions): Let $\emptyset_1, \emptyset_2, \ldots \emptyset_n$ be a sequence of test functions in $\mathcal{D}(\mathbb{R}^n)$. We say that the sequence converges to $\emptyset$ in $\mathcal{D}$ if the following two conditions are satisfied:

a. There is some fixed bounded set $\Omega \subset \mathbb{R}^n$ that contains the support of $\emptyset_1, \emptyset_2, \ldots$ and $\emptyset$.
b. For the derivative of any order α, the sequence converges uniformly to $D^\alpha \emptyset$, i.e.,
   $D^\alpha \emptyset_n \to D^\alpha \emptyset$.

The concept of distribution theory is used to solve problems mathematically where the variation of the function u is highly discontinuous. It can be compared with a device which can measure some very erratic property. Imagine a small test piece for the purpose of measuring some air pollutant. The test piece is inserted in a device, just like a sugar measuring strip, which shows a real number indicating the pollutant. This number is a good indicator of pollution level. I will call the measurement of pollutant in relation with test pieces a distribution. It can be further imagined that the test piece belongs to a well known manufacturer and the test piece is forming a converging sequence. In other words, if the manufacturer of this device increases the thickness of coating on the test piece, the value of indicator converges. If the test pieces are placed at various locations, then the pollutant can be expressed in relation with test pieces via a table $\langle u, \emptyset(x - a_1) \rangle, \langle u, \emptyset(x - a_2) \rangle, \ldots, \langle u, \emptyset(x - a_n) \rangle$. With the help of the test piece, we can measure not only the pollutant concentration but also its rate of change and higher derivatives. Such device will be very useful to study the dynamic behavior of pollutant flow. The device does not exist and it may be difficult to manufacture, but such device will be very helpful to validate many real life problems.

Definition (Distributions): Distributions F on $\mathbb{R}^n$ are a class of functional that maps a set of test functions $\emptyset$ into the set of real numbers $F: \mathcal{D}(\mathbb{R}^n) \to \mathbb{R}$ if

a. $F(a\emptyset + b\psi) = aF(\emptyset) + bF(\psi)$. For every $a, b \in \mathbb{R}$ and $\emptyset, \psi \in \mathcal{D}(\mathbb{R}^n)$.
b. $F(\emptyset_n) \to F(\emptyset)$ (in $\mathbb{R}$) whenever $\emptyset_n \to \emptyset$.

Definition (Regular distribution): A function f is called locally integrable if the integral $\int_K |f(x)| dx$ is finite on every compact $K \subset \Omega$. A distribution $F: D(\Omega) \to \mathbb{R}$ associated with f is called a regular distribution and it can be defined by

$$\langle F, \emptyset \rangle = \int_\Omega f \emptyset \, d\Omega, \quad \emptyset \in D(\Omega) \qquad (5.3)$$

A distribution which is not generated by a locally integrable function is called a *singular distribution*. Dirac delta is an example of a singular distribution because the locally integrable function to generate δ does not exist. Dirac delta distribution is defined by $\langle \delta, \varnothing \rangle = \varnothing(0)$. We cannot interpret $\delta(x)$ as a function in the usual sense but it can be defined only by its action on some test function.

Example 5.2:

The step function $H(x)$, defined by

$$H(x) = \begin{cases} 0 & x < 0 \\ 1 & x \geq 0 \end{cases}$$

generates the distribution H

$$\langle H, \varnothing \rangle = \int_0^\infty H(x)\varnothing(x)\,dx = \int_0^\infty \varnothing(x)\,dx$$

Definition (Distributional and weak derivatives): A function u said to have a kth order distributional derivative v if for every test function $\varnothing \in \mathcal{D}(\Omega)$.

$$\int_\Omega v\varnothing\,dx = (-1)^\alpha \int_\Omega uD^\alpha\varnothing\,dx$$

In other words, v is the kth *generalized derivative* of the regular distribution generated by u.

The function $u \in L^2(\mathbb{R}^n)$ has a *weak derivative* $v = u^k \in L^2(\mathbb{R}^n)$ if

$$\int_\Omega v\varnothing\,d\Omega = (-1)^\alpha \int_\Omega uD^\alpha\varnothing\,d\Omega \ \forall \ \varnothing \in C_0^\infty(\mathbb{R}^n)$$

The *classical derivative* has at least C^0 pointwise continuity on the interval. A weak derivative is a special case of distributional derivative. A weak derivative need only be locally integrable on $\Omega \subset \mathbb{R}^n$.

Example 5.3:

Let

$$f = \begin{cases} 0 & if\ x < 0 \\ x & if\ x \geq 0 \end{cases}$$

Then f is weakly differentiable.

$$\langle f', \varnothing \rangle = -\langle f, \varnothing' \rangle = -\int_{-\infty}^\infty f\varnothing'\,dx = -\int_0^\infty x\varnothing'\,dx = \int_0^\infty x'\varnothing\,dx = \int_0^\infty \varnothing\,dx = \langle H, \varnothing \rangle$$

Therefore, weak derivative of f is the step function H.

Example 5.4:

To find the distributional derivative H' of the Heaviside step function $H(x)$, let us consider the distribution for test function $\varnothing$,

$$\langle H',\varnothing\rangle = -\left\langle H,\frac{d\varnothing}{dx}\right\rangle$$

$$= -\int_0^\infty H(x)\frac{d\varnothing}{dx}dx$$

$$= -\int_0^\infty \frac{d\varnothing}{dx}dx = -[\varnothing]_0^\infty = \varnothing(0) = \langle\delta,\varnothing\rangle$$

So the derivative H' of the step function is the Dirac delta, $H' = \delta$. The step function is not weakly differentiable because $H' \notin L^2(\Omega)$. It is not difficult to prove that the kth distributional derivative of δ is $(-1)^k \varnothing^{(k)}(0)$.

Definition (Weak solution): The function u is not sufficiently differentiable in the classical sense but by a weak solution we mean a function u on $\mathbb{R}^n$ which makes Lu meaningful in the distributional sense. In this case f (equation 5.1) may be a function or distribution.

Example 5.5:

The differential equation

$$x\frac{du}{dx} = 0$$

has a classical solution

$$u(x) = const$$

The differential equation $\frac{du}{dx}$ need not be zero at $x=0$; therefore, we can write this equation as

$$x\frac{du}{dx} = c_1 x\delta(0)$$

or

$$\frac{du}{dx} = c_1\delta(0)$$

Hence,

$$u = c_1 H(x) + c_2$$

This is a weak solution because $H(x)$ is not differentiable in the classical sense at $x = 0$

Definition (Distributional solution): If the function f belongs to the space of distribution then for every $u \in \mathcal{D}'(\mathbb{R}^n)$ (dual of $\mathcal{D}$) satisfying equation 5.1 is called the distributional solution of the equation.

Example 5.6:

Consider a differential equation

$$\frac{d^2 u}{dx^2} = D^2 \delta(x)$$

Since the classical and weak derivative of $\delta(x)$ does not exist, the distributional solution of this equation is

$$u(x) = \delta(x) + c_1 x + c_2$$

5.2 SOBOLEV SPACES

The theory of distribution provides an elegant method of derivative for the functions which are not differentiable in the classical sense. Now it is not difficult to imagine some space for the function with weak derivative. The Sobolev spaces provide a setting for such problems. Sobolev spaces simplify our problem in the sense that the test function need not belong to the space of infinitely differentiable functions. This setting helps in the approximate solution such as the Galerkin and finite element method. Moreover, a special category of Sobolev space is Hilbert space. When dealing with an operator L, in this case a partial differential operator, from space X to another space Y, it is of great value to know beforehand the existence and uniqueness of the solution. Further the Sobolev space is also applicable in variational approach which is a kind of the principle of minimization.

In order to discuss the theory of Sobolev spaces, first we need to discuss the domain. In Sobolev space we will consider an open set in $\mathbb{R}^n$ which is a n-dimensional real Euclidean space. The domains with the following properties are mostly considered in Sobolev spaces: C^m-regularity property, local Lipschitz property, cone property. These properties describe the smoothness of the boundary.

Let $x = (x_1, x_2, ...x_n)$ be a coordinate system to represent the geometry of a boundary. A part of the boundary within a small ball can be expressed as a function

$$x_1 = \xi(x_2, ...x_n).$$

If the function $\xi \in C^m$ for every $x \in \Gamma$, then we say that the function satisfies C^m-regularity property.

A function ξ is Lipschitz continuous at $x \in \Omega \subset \mathbb{R}^n$ if there is a constant c such that

$$\left\| \xi(y) - \xi(x) \right\| \le c \left\| y - x \right\| \tag{5.4}$$

for all $y \in \Omega$ sufficiently near to x. The boundary Γ is said to be Lipschitz if ξ is Lipschitz continuous at the boundary. The domains which contain slit and cusp are examples of non-Lipschitz domain.

For every point $x \in \Omega$ as a vertex, if we can form a cone of some arbitrarily small solid angle and orient it so that all the points of the cone lie in the domain, then we say that the open bounded domain satisfies the cone condition. A three-dimensional cone will look like an ice-cream cone. Domains with cusps will not satisfy the cone condition.

A domain with $C^m (m \ge 1)$ regularity property will satisfy local Lipschitz property. A domain with local Lipschitz property will satisfy cone property. C^m regularity property and local Lipschitz property will be satisfied if the domain lies on one side of its boundary, but this is not essential for cone property. Most of the important property of Sobolev space requires the cone property at the boundary.

5.2.1 $L^p(\Omega)$ Spaces

The $L^p(\Omega)(1 \le p < \infty)$ is a space of all Lebesgue measurable functions $\{u(x)\}$ (Appendix C) on the domain Ω for which

$$\left\| u \right\|_p = \left(\int_{\Omega} |u(x)|^p \, dx \right)^{1/p} < \infty \tag{5.5}$$

The $L^p(\Omega)$ space has a metric

$$d(f, g) = \left\{ \int_{\Omega} |f - g|^p \, dx \right\}^{1/p} \tag{5.6}$$

It is complete, so it is a Banach space. It is a Hilbert space for $p = 2$. $C_0^\infty(\Omega)$ is densely embedded in $L^2(\Omega)$. It means that $f \in L^2(\Omega)$ can sufficiently accurately represent a function $g \in C_0^\infty(\Omega)$.

Definition: The Sobolev space $W^{n,p}(\Omega)$, for ($p \ge 1$), is the space of all functions $u \in L^p(\Omega)$ for which all weak partial derivatives $|\alpha| \le n$ ($n = 0, 1, \ldots\ldots,$) are also elements of $L^p(\Omega)$, i.e.,

$$W^{n,p}(\Omega) = \left\{ u \big| D^\alpha u \in L^p(\Omega), \forall |\alpha| \le n \right\}$$

This is Banach space with the norm:

$$\| u \|_{W^{n,p}(\Omega)} = \left(\int_{\Omega} \sum_{|\alpha| \le n} |D^\alpha u|^p \, dx \right)^{1/p} = \left(\sum_{|\alpha| \le n} \| D^\alpha u \|_{L^p(\Omega)}^p \right)^{1/p}$$

In case of $p=2$ we denote the space $W^{n,2}(\Omega)$ as $H^n(\Omega)$ with the inner product.

$$\left(u,v\right)_{H^n(\Omega)}=(u,v)_n=\iint\sum_{|\alpha|\leq n}\left(D^\alpha u(x)\right)\left(D^\alpha v(x)\right)dx=\sum_{|\alpha|\leq n}\left(D^\alpha u,D^\alpha v\right)_{L^2(\Omega)} \quad (5.7)$$

These Sobolev spaces are Hilbert spaces. It is similar to $C^n(\Omega)$ spaces, but it involves weak derivatives.

$$H^n(\Omega)=\left\{u(x)\,|\,D^\alpha u\in L^2(\Omega),|\alpha|\leq n\right\} \quad (5.8)$$

We define the norm in $H^n(\Omega)\left(\|u\|_n:=\|u\|_{W^{n,2}(\Omega)}\right)$

$$\|u\|_{H^n(\Omega)}=\|u\|_n=\left\{\int_\Omega\sum_{|\alpha|\leq n}\left|D^\alpha u(x)\right|^2 dx\right\}^{1/2} \quad (5.9)$$

5.2.2 Some Common Sobolev Spaces, Norms, and Inner Products

If there is no derivative, then it is the same as that in $L^2(\Omega)$.

$$H^0(\Omega)=L^2(\Omega)=\left\{v(x)\left|\int_\Omega|v|^2\,dx<\infty\right.\right\} \quad (5.10)$$

The inner product in $H^0(\Omega)$ is

$$(u,v)_{H^0(\Omega)}=(u,v)_0=\int_\Omega uv\,dx \quad (5.11)$$

The Sobolev space $H^1(\Omega)$ is defined as

$$H^1(\Omega)=\left\{u(x)\left|\,a<x<b,\int_a^b|u|^2\,dx<\infty,\int_a^b|u'|^2\,dx<\infty\right.\right\} \quad (5.12)$$

The inner product in $H^1(\Omega)$

$$\langle u,v\rangle_{H^1(\Omega)}=\langle u,v\rangle_1=\int_\Omega(uv+u'v')dx \quad (5.13)$$

A norm in $H^1(\Omega)$ can be defined from the inner product

$$\|u\|_1=\int_\Omega\left\{\left(u^2+\left(u'\right)^2\right)dx\right\}^{1/2} \quad (5.14)$$

Similarly for two-dimensional space

$$H^1(\Omega) = \left\{ u(x,y) \,\middle|\, u, \frac{\partial u}{\partial x}, \frac{\partial u}{\partial y} \in L^2(\Omega) \right\} \tag{5.15}$$

The inner product in $H^1(\Omega)$ for two-dimensional space

$$\langle u,v \rangle_{H^1(\Omega)} = \langle u,v \rangle_1 = \iint_\Omega \left(uv + \frac{\partial u}{\partial x}\frac{\partial v}{\partial x} + \frac{\partial u}{\partial y}\frac{\partial v}{\partial y} \right) dxdy \tag{5.16}$$

5.2.3 Relation Between $C^m(\Omega)$ and $H^m(\Omega)$

Sobolev embedding theorems are important to analyze the continuity of function for higher dimensional domain. It is true that a function in $H^1(\Omega)$ is continuous in 1-D but the same is not true for higher dimensional domain.

Let U and V be two Banach spaces, with $U \subseteq V$, and in general, the two spaces are endowed with different norms $\|.\|_V$ and $\|.\|_U$. The space U is called continuously embedded in V if

$$\|u\|_V \leq K \|u\|_U \qquad \forall u \in U \tag{5.17}$$

for some constant $K > 0$ and write $U \hookrightarrow V$. For example $H^1(\Omega) \hookrightarrow L^2(\Omega)$. *We are interested to know under which condition the elements of Sobolev spaces are continuous functions.* This is the content of the following theorem:

The Sobolev Embedding Theorem: Let $\Omega \in \mathbb{R}^n$ be a bounded domain with a Lipschitz boundary. For $m - k > n/2$,

$$H^m(\Omega) \hookrightarrow C^k(\bar{\Omega}) \tag{5.18}$$

In other words if $2m < n$, then

$$H^{m+k} \hookrightarrow C^k, \, k = 0,1,\ldots \tag{5.19}$$

For example, in 1-D space, we have $H^{1+j}(\Omega) \hookrightarrow C^j(\Omega)$ and for 2-D space $H^{2+j}(\Omega) \hookrightarrow C^j(\Omega)$.

Now the question is how to satisfy the boundary condition when u belongs to Sobolev space. The procedure can be explained by introducing an operator called a trace operator. This part is not included in the book but the interested may refer to other books [1–5].

Example 5.7:

Let us consider the problem discussed in Chapter 1. We know that $E\frac{du}{dx}$ is stress in a bar, so $\frac{d}{dx}\left(AE\frac{du}{dx}\right)$ is force per unit length. The Newton principle gives force-displacement equation of a bar:

$$\frac{d}{dx}\left(AE\frac{du}{dx} \right) = -f$$

For simplicity, we assume $AE = 1$ and solve $-\frac{d^2 u}{dx^2} = f$ for continuous force function, i.e., if $f \in C^0(\Omega)$, then the solution is $u = \frac{fx^2}{2} + c_1 x + c_2$. We find constants with the help of boundary conditions and we say $u \in C^2(\Omega) \cap C^0(\bar{\Omega})$, i.e., the function u is two times differentiable and also satisfies the boundary conditions. In terms of mathematical notation, we say $L : C^2(\Omega) \cap C^0(\bar{\Omega}) \to C^0(\Omega)$, where $L = -\frac{d^2 u}{dx^2}$ in this case. In case of many practical problems, the bar has non-uniform cross-section and also is subjected to some concentrated loads, so the equation is converted in the form of energy equation. The finite element formulation utilizes this concept in the form of minimization of potential energy where displacement is expressed in terms of piecewise polynomials. One can understand the reason for this as we calculate the work done due to concentrated load, and "piecewise" strain energy helps to subdivide and assemble the complex domain. But this reason is not sufficient to understand the various solution methodologies of many complex problems. For example, unlike a bar with point load, we cannot solve a problem of two-dimensional membrane which is subjected to point load, i.e., a Poisson's equation with a delta function load. To fix it, we have to use Hilbert and Sobolev spaces.

The embedding theorem shows the relation between these spaces with C^n which contain the function of nth order continuity in the usual sense. Now, instead of looking for a solution in C^n, we are searching a solution in H^n because the continuity conditions in this space are not very strict. The weak continuity of Sobolev space is more generalized and helpful than the energy principles. For bijective mapping, $L : H_B^2 \to H^0$ (H_B^2 contains functions which belong to H^2 space and also satisfies the boundary conditions), we get a unique solution. Here, the eigen functions of the operator L will form the basis functions of the space H_B^2. If the field variable u and force f can be expressed in terms of the basis functions, then we can solve the problem [1]. This approach is used in the collocation method, but it is not suitable to solve the problems like bar with a point load. When we transform the operator L in the energy equation then we get a new operator $I : H_E^1 \to H^{-1}$ (H_E^1 contains functions which belong to H^1 space and satisfies the essential boundary conditions) which we use in FEM, and it is demonstrated in the examples.

Using energy principle, it seems that we can solve the problem of Poisson's equations with point load for two or more dimension also, but before such conclusion we have to analyze the operator $I : H_E^1 \to H^{-1}$. In case of one dimension, u is continuous in H^1, but according to Sobolev inequality theorem, for n-dimensional space, $u(x_1, x_2 \ldots, x_n)$ is continuous in H^s if $s > n/2$. Similarly, $\delta(x) \in H^{-1}$ is continuous but $\delta(x_1, x_2 \ldots, x_n)$ is continuous in H^{-s} if $s > n/2$. Therefore, $\delta(x_1, x_2) \notin H^{-1}$ and correspondingly $u(x_1, x_2) \notin H^1$, so it is inadmissible in the variational problem which is a generalized form of energy equation.

5.3 VARIATIONAL METHOD

This method is developed to find the extremum of functional. Traditionally the problems such as Brachistochrone, isoperimetric and geodesic are solved by the variational method. In the field of solid and structural mechanics the problems are formulated in terms of potential energy (functional), which is required to minimize. Consider the equation

$$-\frac{d^2 u}{dx^2} = f \quad 0 \le x \le L \tag{5.20}$$

The boundary conditions are

$$u(0) = u(L) = 0. \tag{5.21}$$

For a test function w, the equation can be expressed in the alternative form as

$$\int_0^L w\left[-\frac{d^2 u}{dx^2} - f\right] dx = 0 \tag{5.21a}$$

Integration by parts leads to the weak form as

$$\int_0^L \frac{dw}{dx}\frac{du}{dx}dx - \int_0^L wf dx = 0 \tag{5.21b}$$

or

$$a(w,u) - l(w) = 0 \tag{5.22}$$

where $a(w,u)$ represent bilinear form and $l(w)$ is a linear form of the equation. For the symmetric $a(w,u)$, we can get the potential energy form of the equation by substituting $w = \delta u$ (variation in u).

$$\int_0^L \frac{d\delta u}{dx}\frac{du}{dx}dx - \int_0^L \delta u f dx = 0 \tag{5.23a}$$

or

$$\frac{\delta}{2}\int_0^L \left(\frac{du}{dx}\right)^2 dx - \delta\int_0^L uf dx = 0 \tag{5.23b}$$

It is to be noted that δ acts as a differential operator with respect to the dependent variable; therefore, $\delta(u')^2 = 2u'\delta u'$

or

$$\frac{\delta}{2}a(u,u)-\delta l(u)=0=\delta I(u) \tag{5.24}$$

Here $\delta I(u)$ represents the change in potential energy due to variation in u. If we assume variation (or virtual change) of u as $u+\epsilon v$

$$\delta I(u)=\epsilon \int_0^L \lim_{\epsilon \to 0} \frac{I(u+\epsilon v)-I(u)}{\epsilon}dx$$

$$=\epsilon \int_0^L \lim_{\epsilon \to 0} \frac{\left(\frac{1}{2}(u'+\epsilon v')^2-(u+\epsilon v)f\right)-\left(\frac{1}{2}(u')^2-(u)f\right)}{\epsilon}dx$$

$$\tag{5.25}$$

$$=\epsilon \int_0^L \lim_{\epsilon \to 0}[u'v'+\frac{\epsilon}{2}(v')^2+vf]dx$$

$$=\int_0^L \left(u'(\epsilon v')+(\epsilon v)f\right)dx$$

This shows that weak and variational forms of the equation can be used interchangeably. At present the variational form is not limited to the problem of minimization (or maximization) but has wider significance.

The variational problems are stated as: find a function $u \in V$, a Hilbert space, such that

$$a(u,w)=l(w) \quad \forall w \in V. \tag{5.26}$$

Poincaré -Friedrichs inequality, Lax-Milgram lemma and *Céa* lemma are used to check the existence, uniqueness and stability of the solution of elliptic PDEs by variational methods. Let us consider the second order linear operator

$$Lu:=-\sum_{i,j=1}^{d}D_i(a_{ij}D_ju)+\sum_{i=1}^{d}D_i(b_iu)+a_0u \ . \tag{5.27}$$

Its associated bilinear form is

$$a(u,v):=\int_{\Omega}\left[\sum_{i,j=1}^{d}a_{ij}\,D_iu\,D_jv-\sum_{i=1}^{d}b_i\,v\,D_iu+a_0uv\right]. \tag{5.28}$$

Poincaré-Friedrichs inequality: Suppose that Ω is a bounded open set in $\mathbb{R}^n$ (with a sufficiently smooth boundary $\partial\Omega$) and let $u \in H_0^1(\Omega)$; then there exists a constant C_Ω, independent of u, such that

$$\int_\Omega |u|^2 \, dx \leq C_\Omega \sum_{i=1}^n \int_\Omega \left| \frac{\partial u}{\partial x_i} \right|^2 \, dx \tag{5.29}$$

Proof: See [2,6]

Lax-Milgram lemma: Let V be a (real) Hilbert space, endowed with the norm $\|.\|$, $a(u,v) : V \times V \to \mathbb{R}$ be a bilinear form and $l(v) : V \to \mathbb{R}$ a linear continuous functional, i.e., $l \in V'$ where V' denotes the dual space of V. Assume moreover that $a(.,.)$ is continuous, i.e.,

$$\exists \gamma > 0 : |a(u,v)| \leq \gamma \|u\| \|v\| \quad \forall u, v \in V, \tag{5.30}$$

and coercive, i.e.,

$$\exists \alpha > 0 : |a(v,v)| \geq \alpha \|v\|^2 \quad \forall v \in V \tag{5.31}$$

Then, there exists a unique $u \in V$ solution to variational problem, and

$$\|u\| \leq \frac{1}{\alpha} \|l\|_{V'} \tag{5.32}$$

Where the constants α and γ can be expressed as

$$\alpha = \frac{\alpha_0}{1 + C_\Omega}, \tag{5.33}$$

α_0 is ellipticity constant, and

$$\gamma = C_\Omega \left(\max_{1 \leq i,j \leq d} \|a_{ij}\|_{L^\infty(\Omega)} + \|\mathbf{b}\|_{L^\infty(\Omega)} + \|a_0\|_{L^\infty(\Omega)} \right). \tag{5.34}$$

Proof: See [2,6].

The Galerkin approximation to (5.26) read: given $l \in V'$,

find
$$u_h \in V_h : \quad a(u_h, v_h) = l(v_h) \quad \forall v_h \in V_h \tag{5.35}$$

Theorem (*Céa* lemma): Under the assumption of Lax-Milgram lemma there exists a unique solution u_h to (5.35), which furthermore is stable since

$$\|u_h\| \leq \frac{\|l\|_{V'}}{\alpha} \tag{5.36}$$

Moreover, if u is the solution to (5.26), it follows

$$\|u - u_h\| \le \frac{\gamma}{\alpha} \inf_{v_h \in V_h} \|u - v_h\| \tag{5.37}$$

Hence u_h converges to u. Estimate equation (5.37) is usually referred to as *Céa* lemma.

Proof: See [2,6].

We can use the theorems to analyze FEM solution of the following problem:

One of the special cases of the Navier-Stokes equation is the advection-dispersion equation. The search for optimal stabilization methods for advection dominated problems is still an active research area. A simple one-dimensional advection-dispersion equation can be written as

$$-\varepsilon D^2 u + b D u = 0, \; 0 < x < 1. \tag{5.38}$$

For boundary conditions $u(0) = 0$; $u(1) = 1$ with $\varepsilon > 0$ and $b > 0$, the exact solution is

$$u(x) = \frac{e^{bx/\varepsilon} - 1}{e^{b/\varepsilon} - 1}. \tag{5.39}$$

The solution shows a boundary layer (jump) near $x = 1$ if ε / b is small enough. The solution for advection dominated case by finite element, finite difference or finite volume method shows spurious oscillation near jump.

Let us consider again the bilinear form of linear advection-dispersion equation (5.38) to find its coercive and continuity constants.

$$a(u, v) = \int_\Omega \varepsilon \frac{\partial u}{\partial x} \frac{\partial v}{\partial x} dx + \int_\Omega b \frac{\partial u}{\partial x} v \, dx$$

$$\le \left(\varepsilon^2 \int_\Omega \left| \frac{\partial u}{\partial x} \right|^2 dx \int_\Omega \left| \frac{\partial v}{\partial x} \right|^2 dx \right)^{1/2} + \left(C_\Omega b^2 \int_\Omega \left| \frac{\partial u}{\partial x} \right|^2 dx \int_\Omega \left| \frac{\partial v}{\partial x} \right|^2 dx \right)^{1/2} \; [\text{using} (5.29)]$$

$$= \left(\varepsilon + C_\Omega^{1/2} b \right) \left\{ \int_\Omega \left| \frac{\partial u}{\partial x} \right|^2 dx \int_\Omega \left| \frac{\partial v}{\partial x} \right|^2 dx \right\}^{1/2}$$

[Poincaré-Friedrichs inequality]

$$= \left(\varepsilon + C_\Omega^{1/2} b \right) \left\{ \int_\Omega \left| \frac{\partial u}{\partial x} \right|^2 dx \int_\Omega \left| \frac{\partial v}{\partial x} \right|^2 dx \right\}^{1/2}$$

and

$$a(u,u) = \int_\Omega \varepsilon \frac{\partial u}{\partial x} \frac{\partial u}{\partial x} dx + \int_\Omega b \frac{\partial u}{\partial x} u \, dx$$

$$a(u,u) = \int_\Omega \varepsilon \left(\frac{\partial u}{\partial x}\right)^2 dx + \int_\Omega \frac{\partial b}{\partial x} u^2 \, dx$$

$$a(u,u) = \varepsilon \int_\Omega \left(\frac{\partial u}{\partial x}\right)^2 dx$$

Now according to *Céa* lemma

$$\|\nabla u - \nabla u_h\|_0 \le \frac{\gamma}{\alpha} \inf_{v_h \in V_h} \|\nabla u - \nabla v_h\|_0$$

where $\gamma = \varepsilon + C_\Omega^{1/2} b$, $\alpha = \varepsilon$ and if $\Omega = (x_i, x_{i+1}) \subset \mathbb{R}$ then $C_\Omega = \frac{(x_{i+1} - x_i)}{2}$ [6]

The performance of pure Galerkin method is quite poor for large value of γ / ε, i.e., if ε is small in comparison with b. Finite element gives its solution, but it has poor convergence in strain energy due to large value of γ / ε. This is exhibited by the spurious oscillations near the boundary layers. These unphysical oscillations can only be eliminated by severely reducing the mesh-size [7,8] (Figure 5.1).

On the one hand, the theoretical results in (5.39) are true for the convection dominated case, but on the other hand numerical results fail if sufficiently small h is not assumed (which makes discrete problems too expensive or untreatable).

A lot of work has been done for the numerical stability. The first remedy is an "upwinding" scheme which stabilizes computational algorithm but unfortunately introduces numerical "viscosity."

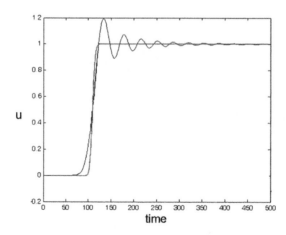

FIGURE 5.1   Spurious oscillations in finite element solution.

Many of the engineering problems do not have this issue. One such example is considered here. The example shows the procedure to find the variational or potential energy form of equation from a differential equation which we have used in Chapter 1 for the derivation of stiffness matrix. The equations of equilibrium for three dimensional continuum, where the Cartesian stress tensor $(\sigma_x, \sigma_y, \sigma_z, \tau_{xy}, \tau_{xz}, \tau_{yz})$ is acting on a finite volume of the solid, can be written as:

$$A(u) = \begin{bmatrix} A_1 \\ A_2 \\ A_3 \end{bmatrix} = -\begin{bmatrix} \dfrac{\partial \sigma_x}{\partial x} + \dfrac{\partial \tau_{xy}}{\partial y} + \dfrac{\partial \tau_{xz}}{\partial z} + b_x \\ \dfrac{\partial \tau_{xy}}{\partial x} + \dfrac{\partial \sigma_y}{\partial y} + \dfrac{\partial \tau_{yz}}{\partial z} + b_y \\ \dfrac{\partial \tau_{xz}}{\partial x} + \dfrac{\partial \tau_{yz}}{\partial y} + \dfrac{\partial \sigma_z}{\partial z} + b_z \end{bmatrix} = 0 \tag{5.40}$$

We assume the weighting function vector as:

$$\delta[u] = [\delta u, \delta v, \delta w]^T$$

The weak form statement can be written as:

$$\int \delta[u]^T A(u)\,d\Omega = -\int_\Omega \left[\delta u(A_1) + \delta v(A_2) + \delta w(A_3)\right]d\Omega = 0 \tag{5.41}$$

Here, we have to use integration by parts, i.e.,

$$\int_\Omega \tau_{12}\frac{\partial u}{\partial x}\,d\Omega = -\int_\Omega \frac{\partial \tau_{12}}{\partial x} u\,d\Omega + \oint_\Gamma n_x u\tau_{12}\,d\Gamma$$

Using it for each term of equation (5.41) and rearranging them, we get

$$\int_\Omega \left[\frac{\partial \delta u}{\partial x}\sigma_x + \left(\frac{\partial \delta u}{\partial y} + \frac{\partial \delta v}{\partial x}\right)\tau_{xy} + \ldots - \delta u b_x - \delta v b_y - \delta w b_z\right]d\Omega$$
$$\tag{5.42}$$
$$-\int_\Gamma \left[\delta u(t_x) + \delta v(t_y) + \delta w(t_z)\right]d\Gamma = 0,$$

where t_x, t_y and t_z are the tractions per unit area acting on external boundary surface.

$$[t] = \begin{bmatrix} t_x \\ t_y \\ t_z \end{bmatrix} = -\begin{bmatrix} n_x\sigma_x + n_y\tau_{xy} + n_z\tau_{xz} \\ n_x\tau_{xy} + n_y\sigma_y + n_z\tau_{yz} \\ n_x\tau_{xz} + n_y\tau_{yz} + n_z\sigma_z \end{bmatrix}. \tag{5.43}$$

Defining virtual strain as

$$\delta[\epsilon] = \begin{bmatrix} \dfrac{\partial \delta u}{\partial x} \\[2mm] \dfrac{\partial \delta v}{\partial y} \\[2mm] \dfrac{\partial \delta w}{\partial z} \\[2mm] \dfrac{\partial \delta u}{\partial y} + \dfrac{\partial \delta v}{\partial x} \\[2mm] \dfrac{\partial \delta v}{\partial z} + \dfrac{\partial \delta w}{\partial y} \\[2mm] \dfrac{\partial \delta w}{\partial x} + \dfrac{\partial \delta u}{\partial z} \end{bmatrix} = \begin{bmatrix} \dfrac{\partial}{\partial x} & 0 & 0 \\[2mm] 0 & \dfrac{\partial}{\partial y} & 0 \\[2mm] 0 & 0 & \dfrac{\partial}{\partial z} \\[2mm] \dfrac{\partial}{\partial y} & \dfrac{\partial}{\partial x} & 0 \\[2mm] 0 & \dfrac{\partial}{\partial z} & \dfrac{\partial}{\partial y} \\[2mm] \dfrac{\partial}{\partial z} & 0 & \dfrac{\partial}{\partial x} \end{bmatrix} \cdot \begin{bmatrix} \delta u \\ \delta v \\ \delta w \end{bmatrix} = [S]\delta[u] \tag{5.44}$$

Equation (5.42) can be written as:

$$\int_\Omega \delta[\epsilon]^T [\sigma] d\Omega - \int_\Omega \delta[u]^T [b] d\Omega - \int_\Gamma \delta[u]^T [t] d\Gamma = 0 \tag{5.45}$$

where $[\sigma]$ is a vector of six stress component and $[b] = [b_x, b_y, b_z]^T$.

In finite element method, we assume field variables as

$$= \begin{bmatrix} N_1 & 0 & 0 & N_2 & 0 & 0 & . & . & . \\ 0 & N_1 & 0 & 0 & N_2 & 0 & . & . & . \\ 0 & 0 & N_1 & 0 & 0 & N_2 & . & . & . \end{bmatrix} \begin{Bmatrix} q_1 \\ q_2 \\ q_3 \\ . \\ . \end{Bmatrix} \tag{5.46}$$

where $N_1, N_2,..$ are the shape functions at nodes 1, 2,.. and $q_1, q_2, q_3...$ are nodal values. We write it as

$$\{u\} = [N]\{q\} \tag{5.47}$$

Substituting it in equation (5.44), we get

$$\delta[\grave{O}] = [S][N]\delta\{q\} = [B]\delta\{q\} \tag{5.48}$$

We have stress-strain relations

$$[\sigma] = [D][\epsilon] \tag{5.49}$$

Substituting equations (5.47) through (5.49) in (5.45), we get

$$\delta\{q\}^T \int_\Omega [B]^T[D][B]d\Omega\{q\} - \delta\{q\}^T \int_\Omega [N]^T[b]d\Omega - \delta\{q\}^T \int_\Gamma [N]^T[t]d\Gamma = 0 \quad (5.50)$$

Since $\delta\{q\}^T$ is arbitrary, we can ignore it and write it in the matrix form as

$$[K]\{q\} = \{f\}$$

This finite element equation is valid for non-linear as well as linear stress-strain relations.

5.4 NONCONFORMING ELEMENTS AND THE PATCH TEST

Convergence of FE solution results in the exact solution of the mathematical model. The criteria for monotonic convergence lie in completeness and compatibility. Completeness requires that the displacement interpolation functions must be chosen such that the elements can represent rigid body motion and constant strain states. Compatibility means the assumed displacement variations are continuous within elements and across inter-element boundaries. It ensures that the strains are bounded within elements and across element boundaries. If displacement is not continuous across element boundaries, then the strains will blow up in-between elements and it leads to erroneous contributions to the potential energy of the structure. The physical meaning is that there are no gaps/cracks which open up when the finite element assemblage is loaded [9].

Monotonic convergence can only be guaranteed for conforming elements, but there are several non-conforming elements (particularly shell elements) in practical use in commercial finite element systems. It has been observed that the results using these elements are often sufficiently accurate for all practical purposes. These nonconforming elements satisfy completeness but do not satisfy compatibility. It can be shown that nonconforming elements result in at least nonmonotonic convergence if the element assemblage as a whole is complete, i.e., they satisfy the patch test. It can be shown that analyses of structures modeled using elements that violate the compatibility requirement will still converge provided that the elements are capable of reproducing a constant strain condition when joined together in an assembly. This can be thought of as an extended completeness condition referring to an assembly of elements, rather than to one element in isolation. It is possible for non-conforming elements that assemblies of elements might be incapable of reproducing the constant strain condition even though the individual elements are complete. Non-conforming elements are therefore required to pass the patch test in which a constant strain field is applied to an assembly of arbitrarily oriented elements. If the element strains represent the constant strain condition, then the test is passed.

Another part of the patch test is to check for strain-free rigid body, both translational and rotational. This involves applying a prescribed displacement to the external nodes of the patch of elements and checking to see that the internal nodes have moved by exactly the same amount and that there are no strains created in any of the elements.

Therefore, one part of the patch test is to check for rigid body motion and another part is to check for constant strain state. The following examples will illustrate the concept in detail:

Example 5.8:

Carry out the patch test of two bar elements as shown in Figure 5.2. For the two elements, the lengths are given as $l_1 = 1$, $l_2 = 4$ cm, uniform cross sectional area $A - 1$ cm^2 and Young's modulus (for both the elements) E is 2×10^5 N/cm^2.

Solution:

The stiffness matrix of the first element is given by

$$k_1 = \frac{EA}{l_1}\begin{bmatrix} 1 & -1 \\ -1 & 1 \end{bmatrix} = 10^5 \begin{bmatrix} 2 & -2 \\ -2 & 2 \end{bmatrix}$$

The stiffness matrix of the second element is given by

$$k_2 = \frac{EA}{l_2}\begin{bmatrix} 1 & -1 \\ -1 & 1 \end{bmatrix} = 10^5 \begin{bmatrix} 0.5 & -0.5 \\ -0.5 & 0.5 \end{bmatrix}$$

Now, the assembled equations of equilibrium is

$$10^5 \begin{bmatrix} 2 & -2 & 0 \\ -2 & 2.5 & -0.5 \\ 0 & -0.5 & 0.5 \end{bmatrix} \begin{Bmatrix} u_I \\ u_{II} \\ u_{III} \end{Bmatrix} = \begin{Bmatrix} f_I \\ f_{II} \\ f_{III} \end{Bmatrix} \tag{5.51}$$

Passing the patch test means satisfying the conditions for rigid body motion and constant strain state.

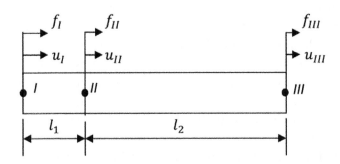

FIGURE 5.2   A patch of two bar elements.

Patch test for rigid body motion:

To test the bar element for rigid body motion, first the displacements of the end nodes are specified to have same value and then solve equation (5.51). Then we see whether the middle node has the same value of displacement or not. Let the end nodal displacements be specified as $u_I = u_{III} = 1$. Then equation (5.1) takes the form

$$10^5 \begin{bmatrix} 2 & -2 & 0 \\ -2 & 2.5 & -0.5 \\ 0 & -0.5 & 0.5 \end{bmatrix} \begin{Bmatrix} 1 \\ u_{II} \\ 1 \end{Bmatrix} = \begin{Bmatrix} R_1 \\ 0 \\ R_3 \end{Bmatrix}$$

where R_1 and R_3 denote the unknown reactions at nodes 1 and 3, respectively. The solution of the second equation is

$$10^5 \left(-2 + 2.5 u_{II} - 0.5 \right) = 0 \text{ or } u_{II} = 1$$

As $u_{II} = 1$, the element passes the patch test for rigid body motion.

Patch test for constant strain:

To test the bar elements for constant strain, the bar elements must undergo a linear displacement. For this we specify the values of the displacements of end nodes as $u_I = 0$ and $u_{III} = 5$. If the displacement variation is linear or if the strain is constant, the displacement of the node 2 must be equal to 1. Using, $u_I = 0$ and $u_{III} = 5$, equation (5.51) takes the form

$$10^5 \begin{bmatrix} 2 & -2 & 0 \\ -2 & 2.5 & -0.5 \\ 0 & -0.5 & 0.5 \end{bmatrix} \begin{Bmatrix} 0 \\ u_{II} \\ 5 \end{Bmatrix} = \begin{Bmatrix} R_1 \\ 0 \\ R_3 \end{Bmatrix}$$

The solution of the second equation is

$$10^5 \left(0 + 2.5 u_{II} - 2.5 \right) = 0 \text{ or } u_{II} = 1$$

As $u_{II} = 1$, the element passes the patch test for constant strain state.

In a similar way we can also illustrate the patch test for two-dimensional elements as shown in Figure 5.3. Here the patch contains four quadrilateral-shaped elements and also consider that the patch is subjected to a uniform stress in the x direction, σ, on the right edge. Now, if the patch test is successful, the finite element solution of the patch should give the stresses as $\sigma_x = \sigma, \sigma_y = 0$, and $\tau_{xy} = 0$ at inside node 5. In a similar spirit, if the patch is subjected to a general stress condition (σ_x, σ_y and τ_{xy} non-zero), we need to conduct patch tests for non-zero constant values of each of the stresses σ_x, σ_y and τ_{xy}, separately.

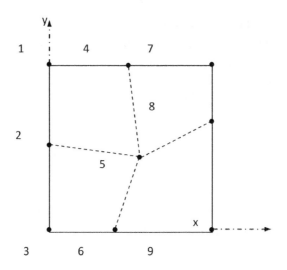

FIGURE 5.3   A patch of quadrilateral elements.

REFERENCES

1. G. Strang and G.J. Fix (1973), *An Analysis of the Finite Element Method*, Prentice-Hall, Englewood Cliffs, NJ.
2. J.N. Reddy (1987), *Applied Functional Analysis and Variational Methods in Engineering*, McGraw-Hill, New York.
3. B.D. Reddy (1998), *Introductory Functional Analysis: With Applications to Boundary Value Problems and Finite Elements*, Springer, New York.
4. J.T. Oden and J.N. Reddy (1976), *An Introduction to the Mathematical Theory of Finite Elements*, Dover Publications, Mineola, New York.
5. J. Chaskalovic (2008), *Finite Element Methods for Engineering Sciences*, Springer-Verlag, Berlin, Germany.
6. A. Quarteroni and A. Valli (2008), *Numerical Approximation of Partial Differential Equations*, Springer, Berlin, Germany.
7. E. Suli (2011), *Finite Element Methods for Partial Differential Equation*, Mathematical Institute, University of Oxford, Oxford, UK.
8. K. Sandeep, S. Gaur, D. Datta and H. S. Kushwaha (2011), "Wavelet based schemes for linear advection–dispersion equation," *Applied Mathematics and Computation*, Vol. 218, No. 7, pp. 3786–3798.
9. S.S. Rao (2004), *The Finite Element Method in Engineering* (4th ed.), Elsevier, Burlington, MA.

CHAPTER 6

Wavelet Based Methods for Differential Equations

T O UNDERSTAND THE CONCEPT of wavelets, first let us recall the concept of Fourier series. As we know, Fourier series are used to approximate a function by using sines and cosines as bases. These bases are non-local, i.e., their supports are from $-\infty$ to $+\infty$ in the time domain. Such long waves are not good in approximating local ripples of a function. A wavelet is a small wave for a finite period with an amplitude that begins at zero, increases, and then dies out with time. These compactly supported bases divide functions into different time and frequency components and therefore are more suited than Fourier methods to approximate choppy signals. There are wide ranges of wavelets available which are useful to cut up mathematical functions into different frequency components. The wavelets are independently developed in the fields of mathematics, quantum physics, engineering, and seismic geology to utilize their specific properties that make them suitable for various applications such as signal processing, image compression, turbulence, earthquake prediction, solution of partial differential equations, etc. The signals or functions which are non-stationary and contain sharp spikes can be represented in terms of a wavelet expansion more appropriately than by using the traditional sine and cosine basis functions.

In Chapter 2, we discussed how wavelet basis functions can be used to solve initial value and boundary value problems. In this chapter, we will discuss the interesting properties of wavelets such as multiscaling, compact support, vanishing moments and orthogonality. Researchers have utilized these properties in various ways for the solution of PDEs. Some researchers have proposed to customize the basis function of FEM. Some have found the wavelet based method to improve the condition number of a large scale finite element matrix. An interesting nonstandard operator for differential and integro-differential equations is developed and used. Wavelet based multigrid methods are also often applied on the iterative solvers which are useful for a large size matrix in the finite element method.

6.1 FUNDAMENTALS OF CONTINUOUS AND DISCRETE WAVELETS

The continuous wavelet transforms (CWT) was introduced by Morlet as a kind of modified windowed transform to analyze geophysical signals. It can be expressed as

$$T(a,b) = \int_{-\infty}^{\infty} f(t)\psi^{*}\left(\frac{t-b}{a}\right) dt \qquad (6.1)$$

In this expression, a is the scaling parameter and b is the translation parameter. The reciprocal of the scaling parameter represents the frequency component and the translation parameter gives the information of time. Superscript * is for the complex conjugate function. In the case of a Haar wavelet, which is a real function, $\psi^{*} = \psi$. Two Haar wavelets are shown in Figure 6.1. The left side Haar wavelet has lower frequency or larger scaling parameter than the right side Haar wavelet.

To understand the convolution process, equation (6.1), consider a box function and a Haar wavelet as shown in Figure 6.2. We flip the Haar wavelet and translate it. For each translation parameter b, calculate the integral 6.1. As we translate the wavelet from the extreme left position to the right end of the signal, at any time b of this translation, some positive part of wavelet overlaps with some positive part of the function and product of the wavelet and the function will be a positive value of the integral. Similarly, overlap of the negative part of the wavelet with the positive part of the signal will result in a negative value of the integral. The convolution of the two functions with translation parameter b is shown in Figure 6.2b.

Consider another example where two functions have the shape of Haar wavelets as shown in Figure 6.3a. The continuous wavelet transform of the function with Haar wavelet is calculated using MATLAB program 6.1 and shown in Figure 6.3b. The convolution gets the highest positive value at a location where both wavelet and part of the signal have the same frequency and same phase. In other words, the positive part of the wavelet overlaps

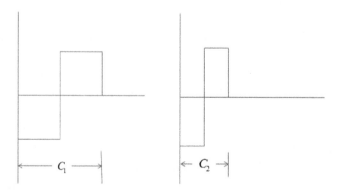

FIGURE 6.1   Haar wavelets with scaling parameter $a = C_1$ and C_2.

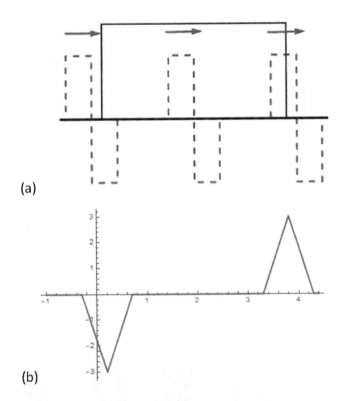

(a)

(b)

FIGURE 6.2 (a) Box function and Haar wavelet at different locations. (b) Convolution of two functions.

with the positive part of the signal and the negative part of the wavelet overlaps with the negative part of the signal. The sum of the two large positive values will be the largest value of the integral. This convolution is plotted in Figure 6.3b along with the scaling parameter (y-axis) and translation parameter (x-axis). Such a convolution plot is used in simultaneous observing the time and frequency of any signal where the highest value indicates the frequency component with time. When the wavelet and the part of the signal have the same frequency but opposite phase or, in other words, the positive part overlaps the negative parts, then the integral will have the largest negative value. Sometimes we show these largest positive and negative values with the help of white and black patches. In the case of different frequencies of signal and wavelets, the perfect overlapping will not occur, so we cannot get the largest positive or negative value. This exercise is also very well presented by Addison [1]. There are few important questions which are not discussed in this chapter. To know more about the accuracy of time and frequency information of the signal and for comparison of continuous wavelet transform with short time Fourier transform, the readers should refer to other books.

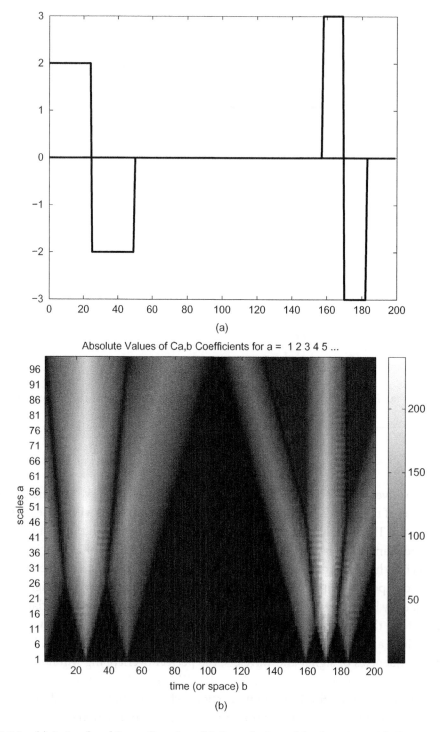

FIGURE 6.3 (a) A simple arbitrary function. (b) Convolution of the function with Haar wavelet.

```
% Matlab program 6.1 for continuous wavelet transform
clc
clear all
t=0:1:199;

for i=1:200
    if i<=25
        u2(i)=2;
    elseif i<=50
        u2(i)=-2;
    else
        u2(i)=0;
    end
end

for i=1:200
    if i<=158
        u3(i)=0;
    elseif i<=170
        u3(i)=3.0;
    elseif i<=183
        u3(i)=-3.0;
    else
        u3(i)=0;
    end
end

x=u2+u3;
%
 plot(t,u2,'r',t,u3,'b','LineWidth',3)

 [psi,xval]=wavefun('haar',10);
 cwt(x,1:100,'haar','plot');
 colormap jet; colorbar;
```

Let us consider a function shown in Figure 6.4a. We want to see both the continuous and discrete wavelet transform of this function. The CWT using Haar wavelet or any other wavelet can be generated as shown in the previous example. For discrete wavelet transform, we express it using basis functions: Haar scaling functions and wavelets. In the case of continuous wavelets, a and b are changed continuously, but for discrete wavelets, the ratio of two successive scaling parameters will be two and dyadic points are used for translation.

As we know that with the help of *discrete wavelet*, we represent any function by using the equation

$$u = \sum_i u_i^0 \phi_i^0 + \sum_k \sum_l d_l^k \psi_i^k \qquad (6.2)$$

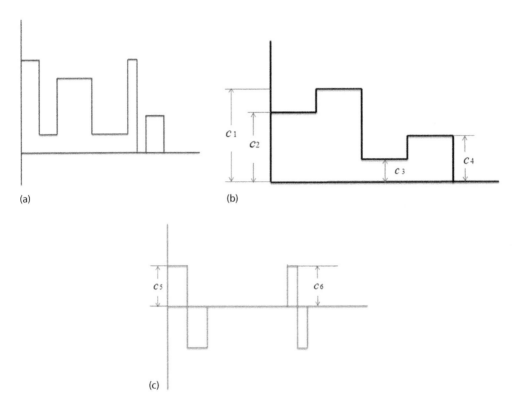

FIGURE 6.4 The signal shown in (a) consists of square waves shown in (b) and (c).

In this particular case, by using Haar scaling function, we can express the square pulse wave as $u_0\phi_0^0 + u_1\phi_1^0 + u_2\phi_2^0 + u_3\phi_3^0$, and it is shown in Figure 6.4b. Here $u_0 = c_2, u_1 = c_1, u_2 = c_3$ and $u_3 = c_4$.

The other part of the function can be expressed by using wavelets as shown in Figure 6.4c:

$$d_0^0\psi_0^0 + 0\psi_1^0 + 0\psi_2^0 + \ldots + 0\psi_0^1 + 0\psi_1^1 + \ldots 0\psi_4^1 + d_5^1\psi_5^1 + 0\psi_6^1$$

where $d_0^0 = -c_5$ and $d_5^1 = -c_6$. Addition of these functions (Figure 6.4b and c) will result (Figure 6.4a).

$$u = u_0\phi_0^0 + u_1\phi_1^0 + u_2\phi_2^0 + u_3\phi_3^0 + d_0^0\psi_0^0 + d_5^1\psi_5^1 \tag{6.3}$$

There are some wavelets whose scaling function ϕ does not exist. For example, the Mexican hat wavelet which is derived from the second derivative function of the Gaussian probability density function and is given as

$$\psi(x) = \left(\frac{2}{\sqrt{3}}\pi^{-\frac{1}{4}}\right) * \left(1 - x^2\right) * e^{-x^2/2} \tag{6.4}$$

The wavelet is shown in Figure 6.5.

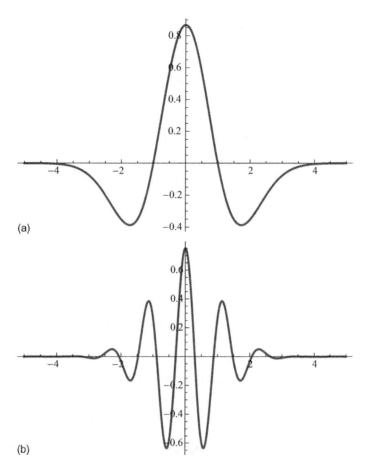

(a)

(b)

FIGURE 6.5   (a) Mexican hat wavelet and (b) Morlet wavelet.

Wavelets like the Mexican hat, Morlet wavelets, etc. are suitable for continuous wavelet transform. In the last example, we assumed a signal which consists of some square pulses, but this can be easily generalized and any function can be discretized and expressed in terms of various scaling and wavelet basis functions. The functions like Figure 6.4a are represented in terms of a coarse approximation, Figure 6.4b along with its fine details Figure 6.4c, and are called multiscaling analysis. Figure 6.4b is a representation of function in a subspace, say $\mathcal{V}^j$. When we use other complementary subspaces of $\mathcal{V}^j$, we find subspaces of details whose functions are shown in Figure 6.4c.

6.2 MULTISCALING

A multiscaling analysis forms a sequence of closed subspaces to satisfy certain self-similarity relations as well as completeness and regularity relations. The subspaces in multiscaling or multi-resolution should satisfy certain conditions.

Multi-resolution analysis of a function space L consists of a sequence of closed subspaces $M = \{ \mathcal{V}^j \subset L | j \in \mathcal{J} \}$ such that

1. $\mathcal{V}^{j-1} \subset \mathcal{V}^j$: To understand this self-similarity relations, consider Figure 2.1a, which shows the function representation in a coarse approximation space $\mathcal{V}^{j-1}$. The function representation in a fine approximation space $\mathcal{V}^j$ is shown in Figure 2.1b. It means that the coarse approximation space is a subspace of fine approximation space. It can be easily proved that the shifted function will also belong to the same space.

2. $U_{j \in \mathcal{J}} V^j$ is dense in L: It means that a function in $\mathcal{V}^j$ can represent a function in L with reasonable accuracy. As shown in Figure 2.1a and b, the union of $\mathcal{V}^j$ is a good approximation of actual space L, and intersection contains zero element $\cap V^j = \{0\}$. This is known as the completeness requirement.

3. The regularity relation demands that the space $\mathcal{V}^j \subset L^2(\mathbb{R})$ has a Reisz basis given by scaling functions $\{ \phi_1^j, \phi_2^j, \phi_3^j \dots, \phi_k^j \dots, \phi_r^j \}$. It means that, if we use basis function ϕ_k^j and its translates, then we can represent any function f with reasonable accuracy. In general, the basic requirement for the basis function is to be piecewise continuous with compact support.

The nestedness $\mathcal{V}^{j-1} \subset \mathcal{V}^j$ is the requirement for multi-resolution, which means that there should exist a refinement relation between ϕ^j (basis function of $\mathcal{V}^j$) and ϕ^{j-1} (basis function of $\mathcal{V}^{j-1}$). As discussed earlier, the relationship has the form

$$\phi_i^{j-1} = \sum_k h_k \phi_{2i+k}^j \tag{6.5}$$

This is the most important fundamental requirement to consider a function as a scaling function in multi-resolution analysis. Every scaling function has unique constants h_i. These constants are called *filter coefficients*. For example, $h_i = \{0.5, 1, 0.5\}$ can be used to generate a b-spline scaling function of the lower resolution as shown in Figure 6.6.

Figure 6.7 shows the uniformly distributed grid points at the different level of resolution used for first generation wavelets transform. It is called a dyadic grid. It is important to notice the change in the grid number 0,1,2,...(first number is 0) with the change in level because this numbering helps when we change function from space $\mathcal{V}^j$ to space $\mathcal{V}^{j-1}$. However, for the second generation wavelets, we can use non-uniform grid points, and the filter coefficients depend on the location.

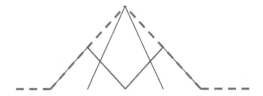

FIGURE 6.6 B-spline scaling function satisfies refinement relation.

FIGURE 6.7 Grid number for spaces $\mathcal{V}^j$, $\mathcal{V}^{j-1}$ and $\mathcal{V}^{j-2}$.

For first generation wavelets, we also have the relation

$$\phi(x) \in \mathcal{V}^j \Leftrightarrow \phi(2x) \in \mathcal{V}^{j+1} \tag{6.6a}$$

$$\phi(x) \in \mathcal{V}^j \Leftrightarrow \phi(x-m) \in \mathcal{V}^j \tag{6.6b}$$

The first relation indicates that for dyadic grid if a basis function belongs to space $\mathcal{V}^{j+1}$ then the basis function which is dilated (expanded) to twice will belong to $\mathcal{V}^j$. According to the second relation, if the basis function belongs to some space, then its translate also belongs to the same space. These relations are valid for all types of first generation scaling functions and wavelets. It means that we can associate each coefficient and basis function with the grid points which are uniformly distributed in the domain. In other words, we are representing the function u^j using constants u_i^j and compactly supported basis functions ϕ_i^j.

The complement of $\mathcal{V}^j$ in $\mathcal{V}^{j+1}$ is defined as subspace $\mathcal{W}^j$ such that

$$\mathcal{V}^{j+1} = \mathcal{V}^j \oplus \mathcal{W}^j \qquad \forall j \in \mathbb{Z} \tag{6.7a}$$

The space $\mathcal{V}^{j+1}$ can be decomposed in a consecutive manner as:

$$\mathcal{V}^{j+1} = \mathcal{V}^0 \oplus \mathcal{W}^0 \oplus \mathcal{W}^1 \oplus \mathcal{W}^2 \dots \oplus \mathcal{W}^j \tag{6.7b}$$

6.3 CLASSIFICATION OF WAVELET BASIS FUNCTIONS

The concept of multiscaling is to represent a function at a coarse level with the help of scaling functions, and any further improvement to the initial approximation is achieved by adding "details" provided by new basis functions known as wavelets.

Each subspace $\mathcal{V}^j$ is spanned by a set of scaling function $\{\phi_i^j(x) | \forall i \in \mathbb{Z}\}$. The basis functions in $\mathcal{W}^j$ are called wavelet functions and are denoted by ψ_i^j. The scaling functions and wavelets can be classified as orthogonal, semi-orthogonal and bi-orthogonal wavelets.

6.3.1 Orthogonal Wavelets

The conditions for orthogonal wavelets are

$$\langle \phi_k^j, \phi_l^j \rangle = \delta_{k,l} \text{ or } \int \phi_k^j \cdot \phi_l^j \, dx = \begin{cases} 1, & \text{for } k=l \\ 0, & \text{for } k \neq l \end{cases} \tag{6.8a}$$

$$\left\langle \psi_k^j , \psi_l^j \right\rangle = \delta_{k,l} \tag{6.8a}$$

$$\left\langle \phi_k^j , \psi_l^j \right\rangle = 0 \tag{6.8b}$$

Examples of orthogonal wavelets are Haar, Daubechies, etc. It is easy to validate these relations using Haar scaling functions and wavelets shown below (Figure 6.8).

The advantage of orthogonal basis functions is again shown here. Consider the representation of a function using orthogonal basis functions

$$u^j = \sum_i u_i^0 \phi_i^0 + \sum_l \sum_{k=0}^{j-1} d_l^k \psi_l^k$$

To find the unknown coefficients u_i^0, both sides of the equation is multiplied with ϕ_k^0 and integrated,

$$\int u^j \phi_k^0 dx = \int \left[\left(\sum_i u_i^0 \phi_i^0 \right) \phi_k^0 + \left(\sum_l \sum_{k=0}^{j-1} d_l^k \psi_l^k \right) \phi_k^0 \right] dx$$

All the terms of the right side integral, except the term associated with $\varphi_{i=k}^0$, vanish because of orthogonal property, and we get

$$\int u^j \phi_k^0 dx = u_k^0$$

Similarly, d_i^k can be obtained because of the orthogonality of ψ_i^j. The method to generate orthogonal basis function is presented in Appendix D. All wavelets are not orthogonal. In such cases, we can get the coefficients to represent any function by using dual.

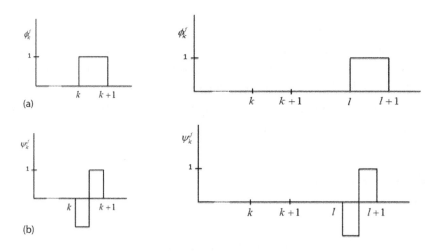

FIGURE 6.8   Translation of (a) Haar scaling functions and (b) Haar wavelets.

6.3.2 Semi-orthogonal Wavelets

In some cases, we sacrifice some orthogonal properties. In this case, wavelets are orthogonal to all coarse scaling function but not orthogonal to each other. In mathematical terms, a semi-orthogonal wavelet basis satisfies the following relations

$$\left\langle \phi_k^j, \phi_l^j \right\rangle \neq \delta_{k,l} \tag{6.9a}$$

$$\left\langle \psi_k^j, \psi_l^j \right\rangle \neq \delta_{k,l} \tag{6.9b}$$

$$\left\langle \phi_k^j, \psi_l^j \right\rangle = 0 \tag{6.9c}$$

6.3.3 Bi-orthogonal Wavelets

Consider a function

$$f(x) = \sum_i c_i \phi_i$$

We have shown that the coefficients c_i can be easily calculated for orthogonal basis functions but if the basis functions are not orthogonal then *there always exist a dual of ϕ_i represented as $\tilde{\phi}_k$ such that*

$$\int f(x) \tilde{\phi}_k dx = \int \left(\sum_i c_i \phi_i \right) \tilde{\phi}_k dx = c_k$$

So, orthogonality is a special case where $\tilde{\phi}_k = \phi_k$

The condition for bi-orthogonality can be written as

$$\left\langle \phi_k^j, \tilde{\phi}_l^j \right\rangle = \delta_{k,l} \tag{6.10a}$$

$$\left\langle \psi_k^j, \tilde{\psi}_l^j \right\rangle = \delta_{k,l} \tag{6.10b}$$

$$\left\langle \phi_k^j, \tilde{\psi}_l^j \right\rangle = 0 \tag{6.10c}$$

$$\left\langle \psi_k^j, \tilde{\phi}_l^j \right\rangle = 0 \tag{6.10d}$$

The method to generate a bi-orthogonal basis function using the lifting scheme is presented in Section 6.6.

6.4 DISCRETE WAVELET TRANSFORM

There are many advantages when we express a function u in terms of scaling function ϕ_k^0 and wavelets ψ_k^j, i.e., in terms of low frequency signals and high frequency signals, respectively. The high frequency signal often contains noise which can be easily filtered, and small detail coefficients are often eliminated to save computer memory. The beautiful part of this method is that we can find the scaling coefficients u_k and detail coefficients d_k^j easily if we discretize u. *The question is how to convert a fine resolution discrete signal into successive coarse resolution and detailed signal efficiently.*

If u^j is a set of discrete values of function u at scale j, then it can be converted into a multi-resolution set u^{j-1} and d^{j-1}, i.e., the coefficients of low frequency and high frequencies.

$$u_i^{j-1} = \int_{-\infty}^{\infty} u \phi_i^{j-1} dx$$

$$= \int_{-\infty}^{\infty} u \left[\frac{1}{\sqrt{2}} \sum_k h_k \phi_{2i+k}^j \right] dx$$

Here $\frac{1}{\sqrt{2}}$ is a part of constant h_k which is used to normalize the energy. We have replaced the basis function of lower resolution space by the basis function of higher resolution space. Rearrangement of the terms gives

$$= \frac{1}{\sqrt{2}} \sum_k h_k \left[\int_{-\infty}^{\infty} u \phi_{2i+k}^j dx \right]$$

$$= \frac{1}{\sqrt{2}} \sum_k h_k u_{2i+k}^j$$

Similarly,

$$d_i^{j-1} = \int_{-\infty}^{\infty} u \psi_i^{j-1} dx = \int_{-\infty}^{\infty} u \left[\frac{1}{\sqrt{2}} \sum_k g_k \phi_{2i+k}^j \right] dx$$

$$= \frac{1}{\sqrt{2}} \sum_k g_k \left[\int_{-\infty}^{\infty} u \phi_{2i+k}^j dx \right] = \frac{1}{\sqrt{2}} \sum_k g_k u_{2i+k}^j$$

Therefore, we have

$$u_i^{j-1} = \frac{1}{\sqrt{2}} \sum_k h_k u_{2i+k}^j \qquad (6.11a)$$

$$d_i^{j-1} = \frac{1}{\sqrt{2}} \sum_k g_k u_{2i+k}^j \qquad (6.11b)$$

These two equations are the most important equations which show that we do not require basis functions for the computation of u_i^{j-1} and d_i^{j-1}. The Mathematica program 6.2

compares the coefficients obtained by both the methods. The discretized function changed into coarse resolution and details using filter coefficients h_k and g_k. This process is known as analysis. The process can be reversed, and it is known as synthesis.

Program 6.2:

MATHEMATICA PROGRAM TO COMPARE THE TWO METHODS FOR CALCULATING COEFFICENTS.
USE OF ORTHOGONAL BASIS FUNCTIONS D4 (db2) AND CONTINUOUS FUNCTION
USE OF FILTER COEFFICIENTS AND DISCRETIZED FUNCTION

```
u = -x^2 + 10 x;

udiscrete = Table[u, {x, 0, 10, 0.5}];

phiD4[x_] = WaveletPhi[DaubechiesWavelet[2], x];

Plot[phiD4[2 x], {x, -0.0, 2.0}, PlotLegends → "Expressions", Exclusions → None];

c = {};

For[i = -2; c1, i < 20, i++,
    c1 = 2 NIntegrate[u * phiD4[2 x - i], {x, 0, 10}, Method → "GaussKronrodRule",
        AccuracyGoal → 6, MaxRecursion → 20, WorkingPrecision → 10]; AppendTo[c, c1];];

u1 = 0; Do[u1 = u1 + c[[i]] * phiD4[2 x + 3 - i], {i, 22}];

Plot[{u, u1}, {x, -0.0, 10.0}, PlotLegends → "Expressions", Exclusions → None];
```

```
WaveletFilterCoefficients[DaubechiesWavelet[2], WorkingPrecision → ∞]
```

$$\left\{\left\{0, \frac{1}{8}\left(1+\sqrt{3}\right)\right\}, \left\{1, \frac{1}{8}\left(3+\sqrt{3}\right)\right\}, \left\{2, \frac{1}{8}\left(3-\sqrt{3}\right)\right\}, \left\{3, \frac{1}{8}\left(1-\sqrt{3}\right)\right\}\right\}$$

```
db4 =
```
$$\left\{\frac{1}{8}\left(1-\sqrt{3}\right), \frac{1}{8}\left(3-\sqrt{3}\right), \frac{1}{8}\left(3+\sqrt{3}\right), \frac{1}{8}\left(1+\sqrt{3}\right)\right\};$$

```
cc = ListConvolve[db4, udiscrete, 1]
 (* Coefficients by using filter coefficients *)
```

```
{5.88321, 1.1875, -0.0707123, 3.06939, 7.5024, 11.4354, 14.8684,
 17.8014, 20.2345, 22.1675, 23.6005, 24.5335, 24.9665, 24.8995,
 24.3325, 23.2655, 21.6986, 19.6316, 17.0646, 13.9976, 10.4306}
```

```
c (* Coefficients by using orthogonal basis functions *)
```

```
{0.008237034015, -0.1588093589, 3.069393305, 7.502406007, 11.43541871,
 14.86843141, 17.80144411, 20.23445681, 22.16746952, 23.60048222, 24.53349492,
 24.96650762, 24.89952032, 24.33253303, 23.26554573, 21.69855843, 19.63157113,
 17.06458383, 13.99759654, 10.43060924, 6.372488609, 1.611394161}
```

It can be observed that the coefficient **c** and **cc** of the Mathematica program are different only at the beginning and end.

Equations (6.11a) and (6.11b) are expressions of convolution for a discretized function with filter coefficients. Similar to the convolution of two continuous functions, the convolution of the two discrete functions $h = \{a, b, c, d\}$ and $u = \{a, b, c, d\}$ can be calculated. The steps are shown below, i.e., flip h, translate, multiply with the function and sum it.

$$u = \{a \quad b \quad c \quad d\}$$
$$\times \ h = \{d \quad c \quad b \quad a\} \qquad\qquad = a^2$$

$$u = \{a \quad b \quad c \quad d\}$$
$$\times \ h = \{d \quad c \quad b \quad a\} \qquad\qquad = 2ab$$

$$u = \{a \quad b \quad c \quad d\}$$
$$\times h = \{d \quad c \quad b \quad a\} \qquad\qquad = b^2 + 2ac$$

$$u = \{a \quad b \quad c \quad d\}$$
$$\times h = \{d \quad c \quad b \quad a\} \qquad\qquad = 2ad + 2bc$$

$$u = \{a \quad b \quad c \quad d\}$$
$$\times h = \{d \quad c \quad b \quad a\} \qquad\qquad = c^2 + 2bd$$

$$u = \{a \quad b \quad c \quad d\}$$
$$\times h = \{d \quad c \quad b \quad a\} \qquad = 2dc$$

$$u = \{a \quad b \quad c \quad d\}$$
$$\times h = \{d \quad c \quad b \quad a\} \quad = d^2$$

Therefore, convolution $h * u = \{a^2, 2ab, b^2 + 2ac, 2ad + 2bc, c^2 + 2bd, 2dc, d^2\}$.

This cumbersome process of finding a discrete convolution can be simplified with the help of the z-transform or Fourier transform. The convolution is a simple multiplication in Fourier and z-transforms; therefore, these two transforms are frequently used for wavelet transform. For example, the z-transform of u and h are $U(z) = a + bz^{-1} + cz^{-2} + dz^{-3}$ and $H(z) = a + bz^{-1} + cz^{-2} + dz^{-3}$, respectively. The convolution of the two functions are

$$H(z).U(z) = \left(a + bz^{-1} + cz^{-2} + dz^{-3}\right).\left(a + bz^{-1} + cz^{-2} + dz^{-3}\right)$$

$$= a^2 + 2abz^{-1} + \left(b^2 + 2ac\right)z^{-2} + \left(2ad + 2bc\right)z^{-3} + \left(c^2 + 2bd\right)z^{-4} + 2dcz^{-5} + d^2z^{-6}$$

For the Fourier transform, we replace z by $e^{j\omega}$. We can also use a matrix in the following way for the convolution

$$
\begin{Bmatrix} 0 \\ c_1 \\ c_2 \\ c_3 \\ c_4 \\ c_5 \\ c_6 \\ c_7 \\ 0 \end{Bmatrix} =
\begin{bmatrix}
h_1 & & & & & & h_4 & h_3 & h_2 & h_1 \\
h_2 & h_1 & & & & & & h_4 & h_3 & h_2 \\
h_3 & h_2 & h_1 & & & & & & h_4 & h_3 \\
h_4 & h_3 & h_2 & h_1 & & & & & & h_4 \\
 & h_4 & h_3 & h_2 & h_1 & & & & & \\
 & & h_4 & h_3 & h_2 & h_1 & & & & \\
 & & & h_4 & h_3 & h_2 & h_1 & & & \\
 & & & & h_4 & h_3 & h_2 & h_1 & &
\end{bmatrix}
\begin{Bmatrix} u_1 \\ u_2 \\ u_3 \\ u_4 \\ 0 \\ 0 \\ 0 \\ 0 \\ 0 \end{Bmatrix}
$$

In the case of convolution for a discrete wavelet transform, both low-pass and high-pass filters are used. Now the question is how to develop an efficient algorithm for discrete wavelet transform. It is always desirable to have filter coefficients h_k and g_k of finite length. The relation between h_k and g_k depends on the orthogonality/bi-orthogonality conditions. To achieve perfect reconstruction of signal, the following equations are used for analysis:

$$
\left\{ \begin{matrix} s^{j-1} \\ d^{j-1} \end{matrix} \right\} = \left[\begin{matrix} A^j \\ B^j \end{matrix} \right] \{ s^j \}. \tag{6.12a}
$$

The synthesis of signal can be expressed as

$$
\{ s^j \} = \left[P^j \mid Q^j \right] \left\{ \begin{matrix} s^{j-1} \\ d^{j-1} \end{matrix} \right\}. \tag{6.12b}
$$

Two-channel filter bank A^j and B^j form an analysis filter bank, whereas P^j and Q^j form a synthesis filter bank [2]. It is often convenient to use the filter bank in z-domain and frequency domain along with up and down sampling. We recommend the book by Strang and Nguyen [3] for further details on the design of filter banks for the readers who are not very familiar with the subject.

The matrix of filter coefficients is used to find the scaling coefficients and detail coefficients. This convolution method of discrete values is not accurate at the both ends of the functions or signal. Therefore, some special techniques such as zero padding or periodicity are used to get the approximate coefficients [4].

Example 6.1:

For various computational purposes, we discretize the functions or the signals. Suppose we have discrete values s_n = {13.2, 13.0, 12.4, 16.6, 24.8, 6.4, 7.8, 8.2} at equal intervals. The process of its analysis is similar to the representation of a function using a Fourier series where we see a function as the sum of its frequency components. Wavelets also split the signal into low and high frequency components. Most of the authors prefer to use Haar wavelet to explain the concept of wavelets because of its simplicity. For analysis, the filter coefficient matrix is used as (equation 6.12a):

$$
\begin{bmatrix}
0.5 & 0.5 & 0 & 0 & 0 & 0 & 0 & 0 \\
0 & 0 & 0.5 & 0.5 & 0 & 0 & 0 & 0 \\
0 & 0 & 0 & 0 & 0.5 & 0.5 & 0 & 0 \\
0 & 0 & 0 & 0 & 0 & 0 & 0.5 & 0.5 \\
-0.5 & 0.5 & 0 & 0 & 0 & 0 & 0 & 0 \\
0 & 0 & -0.5 & 0.5 & 0 & 0 & 0 & 0 \\
0 & 0 & 0 & 0 & -0.5 & 0.5 & 0 & 0 \\
0 & 0 & 0 & 0 & 0 & 0 & -0.5 & 0.5
\end{bmatrix}
\begin{bmatrix}
s_{n,0} \\ s_{n,1} \\ s_{n,2} \\ s_{n,3} \\ s_{n,4} \\ s_{n,5} \\ s_{n,6} \\ s_{n,7}
\end{bmatrix}
=
\begin{bmatrix}
s_{n-1,0} \\ s_{n-1,1} \\ s_{n-1,2} \\ s_{n-1,3} \\ d_{n-1,0} \\ d_{n-1,1} \\ d_{n-1,2} \\ d_{n-1,3}
\end{bmatrix}
\tag{6.13a}
$$

In this example, we get

$$
\begin{bmatrix}
0.5 & 0.5 & 0 & 0 & 0 & 0 & 0 & 0 \\
0 & 0 & 0.5 & 0.5 & 0 & 0 & 0 & 0 \\
0 & 0 & 0 & 0 & 0.5 & 0.5 & 0 & 0 \\
0 & 0 & 0 & 0 & 0 & 0 & 0.5 & 0.5 \\
-0.5 & 0.5 & 0 & 0 & 0 & 0 & 0 & 0 \\
0 & 0 & -0.5 & 0.5 & 0 & 0 & 0 & 0 \\
0 & 0 & 0 & 0 & -0.5 & 0.5 & 0 & 0 \\
0 & 0 & 0 & 0 & 0 & 0 & -0.5 & 0.5
\end{bmatrix}
\begin{bmatrix}
13.2 \\ 13.0 \\ 12.4 \\ 16.6 \\ 24.8 \\ 6.4 \\ 7.8 \\ 8.2
\end{bmatrix}
=
\begin{bmatrix}
13.1 \\ 14.5 \\ 15.6 \\ 8.0 \\ -0.1 \\ 2.1 \\ -9.2 \\ 0.2
\end{bmatrix}
\tag{6.13b}
$$

Now the question is, *what is the advantage of such representation?* The components s_{n-1} show smooth parts of the functions, and the components d_{n-1} are the details. The details are generally low amplitude fluctuations which often represent noise of the system. In other words, we are representing the signal s_n as s_{n-1} = {13.1, 14.5, 15.6, 8.0}, d_{n-1} = {−0.1, 2.1,−9.2,0.2}. Some of the coefficients of d_{n-1} are small e.g.,−0.1 and 0.2.

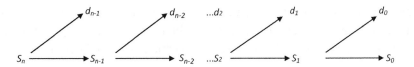

FIGURE 6.9   Analysis of signal.

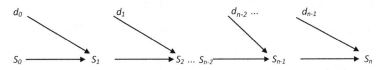

FIGURE 6.10   Synthesis of signal.

Assuming such insignificant detail as zero will save considerable memory without any significant change in the quality of function. During analysis, each successive step generates the average and details.

After ignoring the small details, the signals are reconstructed to their original form. This process is called synthesis. The new synthesized signal is obtained by inverse wavelet transform. The successive analysis and synthesis process are shown in Figures 6.9 and 6.10, respectively. Such analysis and synthesis are extensively used in the field of signal processing, data compression and noise removal.

This concept is also used for the adaptive methods of PDEs solution by many researchers where field variables are converted into average and details, and the insignificant details are ignored. For the solution of a differential equation, we often exploit the multi-resolution property of the well-known Daubechies and b-spline wavelets. However, these wavelets are difficult to apply to the problems with complex domain and boundary conditions. A lifting scheme provides a method to generate some customized wavelets and scaling functions suitable for such problems.

6.5 LIFTING SCHEME FOR DISCRETE WAVELET TRANSFORM

We have seen that a signal can be discretized and transformed to a set of signals which consists of a coarse signal and successive details. The lifting scheme proposed by Sweldens [5] for wavelet transform is more efficient than the methods we discussed in the previous section. The method consists of three simple steps, namely split, predict and update.

1. Split: The first step is to split the discrete signal $\{s_n\}$ into sets of odd and even samples. For example, odd and even sets of a signal which has 2^n entries can be sorted as

$$even_n = \left\{ s_{n,0}, s_{n,2}, s_{n,4}, \ldots s_{n,2^n-2} \right\} \qquad (6.14a)$$

$$odd_n = \left\{ s_{n,1}, s_{n,3}, s_{n,5}, \ldots s_{n,2^n-1} \right\} \qquad (6.14b)$$

2. Predict: In this step, we try to predict odd (or even) indexed samples with the help of even (or odd) indexed samples. Let the predicted set be $P(even_n)$. But there will be some difference between odd_n and $P(even_n)$ which are called detail coefficients

$$\{d_{n-1}\} = odd_n - P(even_n) \tag{6.15}$$

This prediction step is also known as dual lifting.

3. Update: Generally, the next step is the update or primal lifting step where the even set is modified using detail coefficients. The purpose of this step is to maintain the average of coarse scale same as the average of original signal. We can also preserve the moments in the low pass signal. The procedure can be expressed as

$$\{s_{n-1}\} = even_n + U\{d_{n-1}\} \tag{6.16}$$

Sometimes, we normalize or rescale the low pass and high pass signals. There are many wavelet transformations where steps consist of more than one sequence of predict and update and the number of predict and update may be unequal.

Example 6.2:

Suppose we have discrete signal $\{s_3\} = \{13.2, 13.0, 12.4, 16.6, 24.8, 6.4, 7.8, 8.2\}$. We want to represent it in the Haar basis. The lifting scheme can be used as:

Split: $\qquad\qquad\qquad \{even_3\} = \{13.2, 12.4, 24.8, 7.8\}$

$$\{odd_3\} = \{13.0, 16.6, 6.4, 8.2\}$$

We need not create two separate sets while implementing this algorithm because by using proper index numbers, we can write the algorithm for the next operations. Predict: In the case of a Haar transform, it is the difference between odd and even sets.

$$d_{n-1,j} = odd_{n,j} - even_{n,j}$$

$$\{d_{n-1}\} = \{(13.0 - 13.2), (16.6 - 12.4), (6.4 - 24.8), (8.2 - 7.8)\}$$

$$\{d_2\} = \{-0.2, 4.2, -18.4, 0.4\}$$

The detail coefficients are stored in place of the signals of odd index numbers. To make it the same as the previous example, we can rescale it as

$$\{d_2\} = \{-0.1, 2.1, -9.2, 0.2\}$$

Update: The even numbered scaling coefficients are updated to maintain the same average as the original signal. The next level scaling coefficients are calculated using previous signals of even index numbers and the detail coefficients.

$$s_{n-1,j} = even_{n,j} + d_{n-1,j}$$

$$\{s_2\} = \{31.1, 14.5, 15.6, 8.0\}$$

This exercise can be repeated again for $\{s_2\}$. What is the advantage of this exercise? We are splitting signal $\{s_3\}$ in low frequency signal $\{s_2\}$ and high frequency signal $\{d_2\}$. This high frequency component often contains noise which can be easily filtered in this process. It can be observed that $\{d_2\}$ contains small values like -0.1 and 0.2. If we replace these small values with 0 then there will be no significant change in the signal but the binary digits required for these small fluctuations will not be required. This saves a significant amount of computer memory and increases processing time.

6.6 LIFTING SCHEME TO CUSTOMIZE WAVELETS

The lifting scheme approach for discrete wavelet transform, discussed in the last section, is more efficient than convolution of the signal. It can be used for any discrete wavelet transform [6]. The concept of a lifting scheme is also used to customize the bi-orthogonal wavelets. Similar to splitting of the signal, the odd grid points correspond to the detail of the signal and so the associated basis functions are wavelets. Like the predict step of discrete wavelet transform to compute the detail coefficients, the wavelet at any grid point is designed with the help of a higher scaling function seated at that node and the neighboring scaling functions multiplied with some constants. The constants are selected to induce some desired properties like higher vanishing moment, scale orthogonality, etc. Update is the usage of lower scaling functions which satisfies the refinement relations. The even numbered grid point coefficients are the smoother part of the signal. A comparison of discrete wavelet transform and customization of wavelet is shown in Table 6.1.

The customized wavelets can be easily generated using Mathematica programs. In programs 6.4 and 6.5, two different approaches are used to generate the wavelets for the hat functions as the scaling functions.

We have demonstrated with the help of program 6.2 that the coefficients $\{u_i^j, d_i^j\}$ can be obtained using the orthogonal basis functions ϕ_k^j and ψ_i^j as well as by using high and low pass filter coefficients $\{h_i, g_i\}$. Prof. Gilbert Strang [3] called these filter coefficients "magic numbers," and we will show that these numbers can be used to generate the basis functions. With the help of the subdivision scheme, the scaling function is generated by using low-pass filters and subsequently the mother wavelet by using the high-pass filters in the Mathematica program 6.3.

TABLE 6.1   Lifting Scheme for Discrete Wavelet Transform and Customized Wavelets

Steps	Discrete wavelet transform: Initial Signal:	Customized bi-orthogonal wavelet: Initial Basis Function:
	$\{s_n\} = \left\{ s_{n,0}, s_{n,1}, s_{n,2}, \ldots s_{n,2^n-1} \right\}$	$\{\phi_0^{j+1}, \phi_1^{j+1}, \phi_2^{j+1}, \ldots, \phi_{2n}^{j+1}\}$
Split	odd_n and $even_n$	$\{\phi_1^{j+1}, \phi_3^{j+1}, \ldots, \phi_{2n-1}^{j+1}\}$ and $\{\phi_0^{j+1}, \phi_2^{j+1}, \ldots, \phi_{2n}^{j+1}\}$
Predict	$\{d_{n-1}\} = odd_n - P(even_n)$	$\psi_k^j = \phi_{2k-1}^{j+1} - \sum_i S_i \phi_i^j$
Update	$\{s_{n-1}\} = even_n + U\{d_{n-1}\}$	$\phi_i^j = \sum_k h_k \phi_{2i+k}^{j+1}$

Program 6.3:

Subdivision Scheme for Primal Scaling Functions and Wavelets

$h0 = \dfrac{1}{2}$; $h1 = 1$; $h2 = \dfrac{1}{2}$;

$g0 = \dfrac{1}{516}\ 19$; $g1 = \dfrac{1}{516}\ (38)$; $g2 = \dfrac{1}{516}\ (-62)$; $g3 = \dfrac{1}{516}\ (-162)$; $g4 = \dfrac{1}{516}\ (354)$;

$g5 = \dfrac{1}{516}\ (-162)$; $g6 = \dfrac{1}{516}\ (-62)$; $g7 = \dfrac{1}{516}\ (38)$; $g8 = \dfrac{1}{516}\ (19)$;

$\text{pphi0}[t\_] = \begin{cases} 1 & 0 \le t < 1 \\ 0 & t \ge 0 \end{cases}$

$\begin{cases} 1 & 0 \le t < 1 \\ 0 & \text{True} \end{cases}$

`pphi1[t_] = h0 * pphi0[2 t] + h1 * pphi0[2 t - 1] + h2 * pphi0[2 t - 2];`

`Plot[pphi0[t], {t, -0.2, 2.0}]`

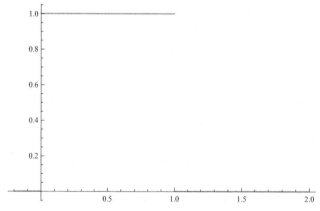

`pphi2[t_] = h0 * pphi1[2 t] + h1 * pphi1[2 t - 1] + h2 * pphi1[2 t - 2];`

`pphi3[t_] = h0 * pphi2[2 t] + h1 * pphi2[2 t - 1] + h2 * pphi2[2 t - 2];`

`Plot[pphi3[t], {t, -0.0, 3.0}]`

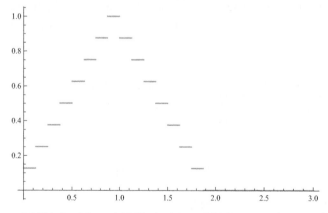

`pphi4[t_] = h0 * pphi3[2 t] + h1 * pphi3[2 t - 1] + h2 * pphi3[2 t - 2];`

```
Plot[pphi4[t], {t, -0.0, 3.0}]
```

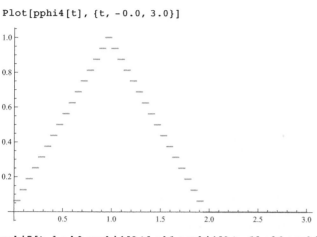

```
pphi5[t_] = h0 * pphi4[2 t] + h1 * pphi4[2 t - 1] + h2 * pphi4[2 t - 2];
```

```
Plot[pphi5[t], {t, -0.0, 3.0}]
```

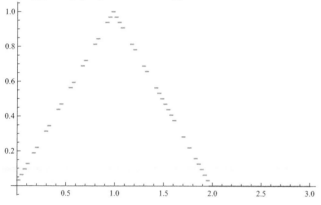

```
pphi6[t_] = h0 * pphi5[2 t] + h1 * pphi5[2 t - 1] + h2 * pphi5[2 t - 2];
```

```
Plot[pphi6[t], {t, -0.0, 3.0}]
```

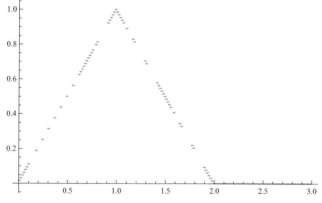

```
pphi7[t_] = h0 * pphi6[2 t] + h1 * pphi6[2 t - 1] + h2 * pphi6[2 t - 2];
```

```
Plot[pphi7[t], {t, -0.0, 3.0}]
```

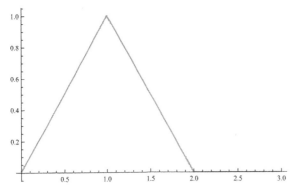

```
ppsi[t_] = g0 * pphi6[2 t - 4] + g1 * pphi6[2 t - 5] + g2 * pphi6[2 t - 6] +
    g3 * pphi6[2 t - 7] + g4 * pphi6[2 t - 8] + g5 * pphi6[2 t - 9] +
    g6 * pphi6[2 t - 10] + g7 * pphi6[2 t - 11] + g8 * pphi6[2 t - 12];
```

```
Plot[ppsi[t], {t, 0.0, 8.0}, PlotRange → All]
```

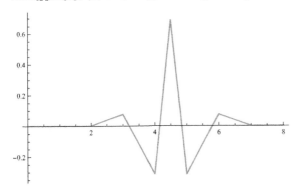

Program 6.4:
Program for Customized Wavelets

$$pphi[t\_] = \begin{cases} t & 0 \le t \le 1 \\ 2 - t & 1 \le t \le 2 \end{cases}$$

$$\begin{cases} t & 0 \le t \le 1 \\ 2 - t & 1 \le t \le 2 \\ 0 & True \end{cases}$$

```
Plot[pphi[t], {t, -3, 3.0}]
```

```
Plot[{pphi[2 t], pphi[t - 1], pphi[t], pphi[t + 1], pphi[t + 2]},
    {t, -2, 5}, PlotRange → All]
```

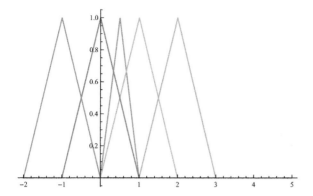

```
ppsi[t_] = pphi[2 t] - (s1 * pphi[t - 1] + s2 * pphi[t] + s3 * pphi[t + 1] + s4 * pphi[t + 2]);
```

$$\int_{-6}^{6} \text{ppsi}[t]\, dt$$

$$\frac{1}{2}\,(1 - 2\, s1 - 2\, s2 - 2\, s3 - 2\, s4)$$

$$\int_{-6}^{5} \text{ppsi}[t] * t\, dt$$

$$\frac{1}{4}\,(1 - 8\, s1 - 4\, s2 + 4\, s4)$$

$$\int_{-5}^{5} \text{ppsi}[t] * t{}^{\wedge}2\, dt$$

$$\frac{1}{48}\,(7 - 200\, s1 - 56\, s2 - 8\, s3 - 56\, s4)$$

$$\int_{-5}^{5} \text{ppsi}[t] * t{}^{\wedge}3\, dt$$

$$-\frac{3}{32}\,(-1 + 96\, s1 + 16\, s2 - 16\, s4)$$

```
LinearSolve[{{-1, -1, -1, -1}, {-8, -4, 0, 4},
   {-200, -56, -8, -56}, {96, 16, 0, -16}}, {-1 / 2, -1, -7, 1}]
```

$$\left\{-\frac{3}{64}, \frac{19}{64}, \frac{19}{64}, -\frac{3}{64}\right\}$$

```
ppsi[t_] =
```

$$\text{pphi}[2\, t] - \left(-\frac{3}{64} * \text{pphi}[t - 1] + \frac{19}{64} * \text{pphi}[t] + \frac{19}{64} * \text{pphi}[t + 1] - \frac{3}{64} * \text{pphi}[t + 2]\right);$$

```
Plot[ppsi[t], {t, -3.0, 4.0}, PlotRange → All]
```

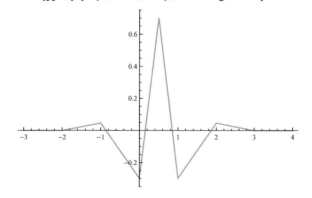

Program 6.5:

Program for Customized Wavelets

$$\texttt{pphi[t\_]} = \begin{cases} t & 0 \le t \le 1 \\ 2-t & 1 \le t \le 2 \end{cases}$$

$$\begin{cases} t & 0 \le t \le 1 \\ 2-t & 1 \le t \le 2 \\ 0 & \text{True} \end{cases}$$

```
Plot[{pphi[2 t], pphi[t - 1], pphi[t], pphi[t + 1], pphi[t + 2]},
  {t, -2, 5}, PlotRange → All]
```

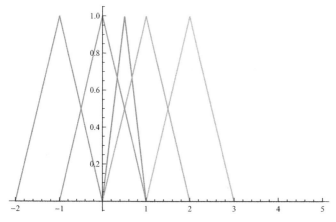

```
ppsi[t_] = pphi[2 t] - (s1 * pphi[t - 1] + s2 * pphi[t] + s3 * pphi[t + 1] + s4 * pphi[t + 2]);
```

$$\int_{-6}^{6} \texttt{ppsi[t]} * \texttt{pphi[t - 1]} \, dt$$

$$\frac{1}{6}(-4\,s1 - s2)$$

$$\int_{-6}^{5} \texttt{ppsi[t]} * \texttt{pphi[t]} \, dt$$

$$\frac{1}{12}(3 - 2\,s1 - 8\,s2 - 2\,s3)$$

$$\int_{-5}^{5} \texttt{ppsi[t]} * \texttt{pphi[t + 1]} \, dt$$

$$\frac{1}{12}(3 - 2\,s2 - 8\,s3 - 2\,s4)$$

$$\int_{-5}^{5} \texttt{ppsi[t]} * \texttt{pphi[t + 2]} \, dt$$

$$\frac{1}{6}(-s3 - 4\,s4)$$

```
LinearSolve[
 {{-4 , -1, 0, 0}, {-2 , -8 , -2 , 0}, {0, -2 , -8 , -2 }, {0, 0, -1, -4 }}, {0, -3, -3, 0}]
```

$$\left\{-\frac{3}{38}, \frac{6}{19}, \frac{6}{19}, -\frac{3}{38}\right\}$$

```
ppsi[t_] =
```

$$\texttt{pphi[2 t] -} \left(-\frac{3}{38} * \texttt{pphi[t - 1] +} \frac{6}{19} * \texttt{pphi[t] +} \frac{6}{19} * \texttt{pphi[t + 1] -} \frac{3}{38} * \texttt{pphi[t + 2]}\right);$$

```
Plot[ppsi[t], {t, -3.0, 4.0}, PlotRange → All]
```

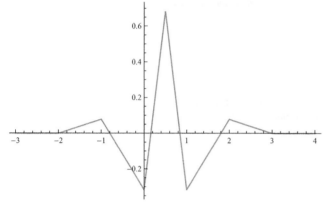

Programs 6.4 and 6.5 show the implementation of a lifting scheme for customization of wavelets. In these cases, we used simple hat functions as the scaling function. However, more complex scaling functions can be used and customized wavelets can be generated.

For perfect reconstruction, the primal and duals of the bi-orthogonal filter bank should satisfy the following relations [3]

$$\tilde{h}(z)h(-z)+\tilde{g}(z)g(-z)=0 \tag{6.17a}$$

$$\tilde{h}(z)h(z)+\tilde{g}(z)g(z)=2z^{l} \tag{6.17b}$$

If we assume $\tilde{h}(z)=g(-z)$ and $\tilde{g}(z)=-h(-z)$, equation (6.18a) will be satisfied. For the known filters $h(z)=h_0 + h_1 z^{-1} + h_2 z^{-2} +\dots$ and $g(z)=g_0 + g_1 z^{-1} + g_2 z^{-2} +\dots$ we can find the duals $\tilde{h}(z)=\tilde{h}_0 + \tilde{h}_1 z^{-1} + \tilde{h}_2 z^{-2} +\dots$ and $\tilde{g}(z)=\tilde{g}_0 + \tilde{g}_1 z^{-1} + \tilde{g}_2 z^{-2} +\dots$.

Example 6.3:
$h = \{0.5,1,0.5\}$ and $g=\left\{\frac{-3}{38}, \frac{6}{19}, \frac{6}{19}, \frac{-3}{38}\right\}$, then we get $\tilde{g}(z)=\{-0.5,1,-0.5\}$ and $\tilde{h}(z)=\left\{\frac{-3}{38}, \frac{-6}{19}, \frac{6}{19}, \frac{3}{38}\right\}$. It can be tested that it is satisfying equation (6.17).

The dual basis functions can be generated with the help of a subdivision scheme using the corresponding filter coefficients. Similar to the Mathematica program 6.3 for primal basis function, the Mathematica program 6.6 generates the dual basis functions.

Program 6.6:

Dual Scaling Functions and Wavelets

```
h0 = 19 / 516; h1 = -38 / 516; h2 = -62 / 516; h3 = 162 / 516;
h4 = 354 / 516; h5 = 162 / 516; h6 = -62 / 516; h7 = -38 / 516; h8 = 19 / 516;

g0 = -1 / 2; g1 = 1; g2 = -1 / 2;

pphi0[t_] = {  1   0 ≤ t < 1
               0   t ≥ 1

{  1   0 ≤ t < 1
   0   True

pphi1[t_] = h0 * pphi0[2 t] + h1 * pphi0[2 t - 1] + h2 * pphi0[2 t - 2] +
    h3 * pphi0[2 t - 3] + h4 * pphi0[2 t - 4] + h5 * pphi0[2 t - 5] +
    h6 * pphi0[2 t - 6] + h7 * pphi0[2 t - 7] + h8 * pphi0[2 t - 8];

Plot[pphi0[t], {t, -2, 2.0}]
```

```
pphi2[t_] = h0 * pphi1[2 t] + h1 * pphi1[2 t - 1] + h2 * pphi1[2 t - 2] +
    h3 * pphi1[2 t - 3] + h4 * pphi1[2 t - 4] + h5 * pphi1[2 t - 5] +
    h6 * pphi1[2 t - 6] + h7 * pphi1[2 t - 7] + h8 * pphi1[2 t - 8];

pphi3[t_] = h0 * pphi2[2 t] + h1 * pphi2[2 t - 1] + h2 * pphi2[2 t - 2] +
    h3 * pphi2[2 t - 3] + h4 * pphi2[2 t - 4] + h5 * pphi2[2 t - 5] +
    h6 * pphi2[2 t - 6] + h7 * pphi2[2 t - 7] + h8 * pphi2[2 t - 8];

pphi4[t_] = h0 * pphi3[2 t] + h1 * pphi3[2 t - 1] + h2 * pphi3[2 t - 2] +
    h3 * pphi3[2 t - 3] + h4 * pphi3[2 t - 4] + h5 * pphi3[2 t - 5] +
    h6 * pphi3[2 t - 6] + h7 * pphi3[2 t - 7] + h8 * pphi3[2 t - 8];
```

```
Plot[pphi4[t], {t, -2, 8}, PlotRange → All]
```

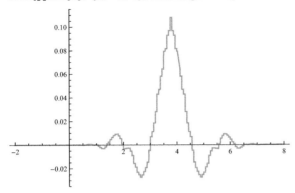

```
pphi5[t_] = h0 * pphi4[2 t] + h1 * pphi4[2 t - 1] + h2 * pphi4[2 t - 2] +
    h3 * pphi4[2 t - 3] + h4 * pphi4[2 t - 4] + h5 * pphi4[2 t - 5] +
    h6 * pphi4[2 t - 6] + h7 * pphi4[2 t - 7] + h8 * pphi4[2 t - 8];
```

```
Plot[pphi5[t], {t, -2, 8}, PlotRange → All]
```

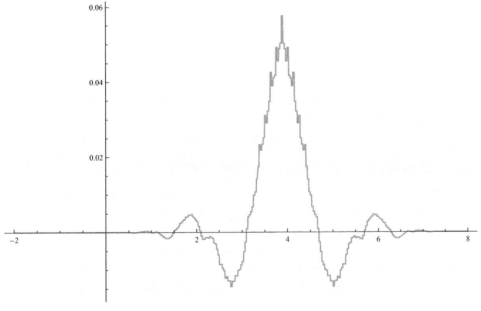

```
ppsi[t_] = g0 * pphi5[2 t] + g1 * pphi5[2 t - 1] + g2 * pphi5[2 t - 2];
```

`Plot[ppsi[t], {t, -3.0, 8.0}, PlotRange → All]`

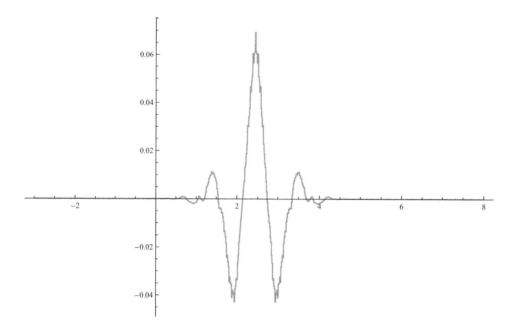

In Chapter 2, we solved some PDEs using the wavelet bases. Most of the available wavelets provide stable Riesz basis, and the methods have better condition numbers than FEM or FDM. But in practical applications, wavelet methods are not yet ready to compete with the traditional FEM approach. It is difficult to use first generation wavelets such as Daubechies wavelets in complex domain of engineering problems. The combination of wavelets with FDM, FEM and SEM can show good results. Heedene [7], Sudarshan [8], and Amaratunga and Sudarshan [9] have customized wavelets using a lifting scheme. In this section, we are presenting some simple cases for solution of differential equations.

We generally use discretization of domain and shape function for FEM. The one-dimensional case is shown in Figure 6.11a. The same can be expressed in terms of basis functions as shown in Figure 6.11b though it may not be convenient to find the element stiffness matrix.

Consider a one-dimensional field variable $u = u_0^j \phi_0^j + u_1^j \phi_1^j + u_2^j \phi_2^j \ldots + u_n^j \phi_n^j$ expressed in terms of the basis function $\{\phi_0^j, \phi_1^j, \phi_2^j, \ldots, \phi_n^j\}$ which satisfies the refinement relations. As usual, the nodal values $\{u_0^j, u_1^j, u_2^j, \ldots, u_n^j\}$ are computed by solving the

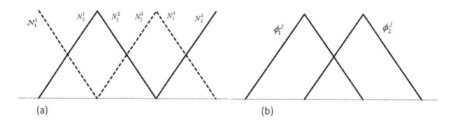

FIGURE 6.11 (a) Finite element shape functions and (b) equivalent hat basis functions.

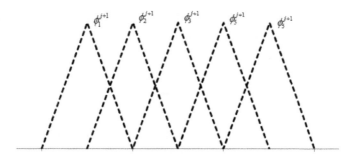

FIGURE 6.12   Higher scaling functions for refinement (or increased FE nodes).

global stiffness matrix. To improve the solution, new nodes are added between the two successive nodes of the initial mesh. The modified field variable can be expressed as $u = u_0^{j+1}\phi_0^{j+1} + u_1^{j+1}\phi_1^{j+1} + u_2^{j+1}\phi_2^{j+1}\ldots + u_{2n}^{j+1}\phi_{2n}^{j+1}$. The nodes and the basis functions are shown in Figure 6.12

Since there exists a refinement relation between the two scales, we replace the even numbered basis functions with the previous scale basis functions. In other words, without changing the basis functions of the previous scale, we add the higher scale basis function at the new nodes. This is known as hierarchical basis method and is shown in Figure 6.13. The field variable can be defined as

$$u = u_0\phi_0^j + u_1\phi_1^{j+1} + u_2\phi_1^j + u_3\phi_3^{j+1} + u_4\phi_2^j + \ldots + u_{2n-1}\phi_{2n-1}^{j+1} + u_{2n}\phi_n^j$$

The hierarchical basis $\{\phi_1^{j+1}, \phi_3^{j+1}, \ldots, \phi_{2n-1}^{j+1}\}$ are also called lazy wavelets. These wavelets can be modified using a lifting scheme to improve the condition number of global stiffness matrix and in some cases, we can get scale orthogonal stiffness matrix [10].

This method can be applied with any finite element basis function which satisfies the refinement relations

$$\phi_k^j = \sum_l h_{k,l}\phi_l^{j+1}. \tag{6.18a}$$

The grid points need not be uniformly distributed. This is suitable for finite element structure because the field variables are generally not expressed by uniformly distributed

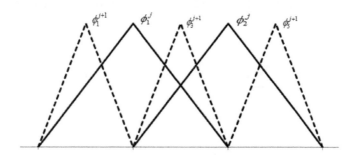

FIGURE 6.13   Hierarchical basis method.

basis functions. In this case, the filter coefficients $h_{k,l}$ depend on the location of basis functions. Wavelets are generated by lifting scheme

$$\psi_k^j = \phi_{2k-1}^{j+1} - \sum_i S_i \phi_i^j \tag{6.18b}$$

Application of equation (6.18a) and (6.18b) in variational method will generate a multi-resolution stiffness matrix similar to equation (2.16) with coefficients.

$$k_{il} = \int_\Omega \phi_i A \phi_l d\Omega, \tag{6.19a}$$

$$\gamma_{il} = \int_\Omega \psi_i A \psi_l d\Omega, \tag{6.19b}$$

$$\alpha_{il} = \int_\Omega \phi_i A \psi_l d\Omega, \tag{6.19c}$$

$$\beta_{il} = \int_\Omega \psi_l A \phi_i d\Omega, \tag{6.19d}$$

$$f_i = \int_\Omega \phi_i f d\Omega, \tag{6.19e}$$

$$g_i = \int_\Omega \psi_i f d\Omega. \tag{6.19f}$$

Sudarshan [8] applied a lifting scheme on some finite element basis function to develop scale orthogonal wavelets which decouples the refinement process. He developed an efficient algorithm to find filter coefficients S_i which generates scale orthogonal wavelets, i.e., the coupling terms will be eliminated $\alpha_{il} = \beta_{il} = 0$. In this case, equation (2.16) can be written as:

$$\sum_{l=1}^n k_{il} c_l = f_i \qquad \text{for } i = 1, \dots n \tag{6.20a}$$

$$\sum_{l=1}^n \gamma_{il} d_l = g_i \qquad \text{for } i = 1, \dots n \tag{6.20b}$$

These equations gives coarse solution and refinement, and can be solved independently. Due to finite element basis functions, the method is suitable for complex domain problems. The boundary conditions can be easily applied.

Mathematica program 6.6 shows customized wavelets for cubic Hermite basis function which is used by Sudarshan [11] to improve the condition number of stiffness matrix. The scale orthogonal wavelets are used in program 6.7.

Example 6.4:

In the case of a beam, each node has displacement and rotational degrees of freedom, denoted as u and θ, respectively. The scaling functions (or finite element shape functions) of both displacement and rotation must satisfy refinement relations. The vector form of such relation can be expressed as

$$\begin{Bmatrix} \phi_{j,k}^u(x) \\ \phi_{j,k}^\theta(x) \end{Bmatrix} = \begin{Bmatrix} \phi_{j+1,k}^u(x) \\ \phi_{j+1,k}^\theta(x) \end{Bmatrix} + \sum_m H_j[k,m] \begin{Bmatrix} \phi_{j+1,m}^u(x) \\ \phi_{j+1,m}^\theta(x) \end{Bmatrix} \tag{6.21a}$$

where m are the subdivision points. It can be verified that the cubic Hermite basis function satisfies equation (6.21a) for the following matrix $H_j[k,m]$:

$$H_j[k,m] = \begin{bmatrix} \phi_{j,k}^u(x_m) & \dfrac{d\phi_{j,k}^u(x_m)}{dx} \\ \phi_{j,k}^\theta(x_m) & \dfrac{d\phi_{j,k}^\theta(x_m)}{dx} \end{bmatrix} \tag{6.21b}$$

We can use a lifting scheme to customize the wavelets. It is called multiwavelets because the nodes have more than one degree of freedom. The predict step can be expressed as

$$\begin{Bmatrix} w_{j,m}^u(x) \\ w_{j,m}^\theta(x) \end{Bmatrix} = \begin{Bmatrix} \phi_{j+1,m}^u(x) \\ \phi_{j+1,m}^\theta(x) \end{Bmatrix} - \sum_k S_j^T[k,m] \begin{Bmatrix} \phi_{j+1,k}^u(x) \\ \phi_{j+1,k}^\theta(x) \end{Bmatrix} \tag{6.22}$$

Figure 6.14 shows the grid at level $j + 1$ for scaling functions and corresponding grid at level j for Hermite interpolating scaling functions and wavelets over the domain.

The basis function of the Euler beam can be expressed as [12]

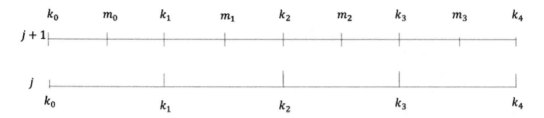

FIGURE 6.14   Grid at level j and $j + 1$.

$$\phi^u(x) = \begin{cases} 1 - 3\left(\dfrac{x}{h_e}\right)^2 + 2\left(\dfrac{x}{h_e}\right)^3 & 0 \leq x \leq h_e \\[3mm] 1 - 3\left(\dfrac{x}{h_e}\right)^2 - 2\left(\dfrac{x}{h_e}\right)^3 & -h_e \leq x \leq 0 \\[3mm] 0 & Elsewhere \end{cases} \tag{6.23a}$$

$$\phi^\theta(x) = \begin{cases} x\left(1 - \dfrac{x}{h_e}\right)^2 & 0 \leq x \leq h_e \\[3mm] x\left(1 + \dfrac{x}{h_e}\right)^2 & -h_e \leq x \leq 0 \\[3mm] 0 & Elsewhere \end{cases} \tag{6.23b}$$

where h_e is length of the element. In this case, we get the matrix as:

$$H_j[k,m] = \begin{bmatrix} 0.5 & 0.15 \\ -0.625 & -0.125 \end{bmatrix}$$

The Mathematica program 6.7 shows that the Hermite scaling functions satisfy the refinement relations and program 6.8 uses a lifting scheme to generate the Hermite wavelet. The condition number of the Euler beam stiffness matrix reduces if customized wavelets along with scaling function are used.

Program 6.7:

(* MATHEMATICA PROGRAM 6.7 *)

```
(* THIS PROGRAM SHOWS THAT THE HERMITE
   SCALING FUNCTIONS SATISIFIES REFINEMENT RELATIONS *)
he = 5;
```

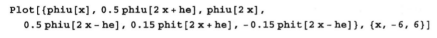

```
dphiu[x_] = D[phiu[x], x];

Plot[{phiu[x]}, {x, -6, 6}];

Plot[{phiu[2 x]}, {x, -6, 6}];
```

$$\text{phit}[x\_] = \begin{cases} x\,(1-x/he)^2 & 0 \le x \le he \\ x\,(1+x/he)^2 & -he \le x \le 0 \end{cases};$$

```
Plot[{phiu[x], 0.5 phiu[2 x + he], phiu[2 x],
  0.5 phiu[2 x - he], 0.15 phit[2 x + he], -0.15 phit[2 x - he]}, {x, -6, 6}]
```

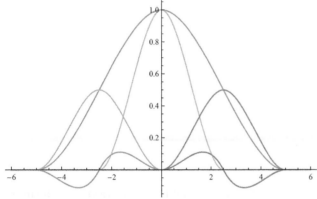

```
phiun[x_] = 0.5 phiu[2 x + he] + phiu[2 x] +
  0.5 phiu[2 x - he] + 0.15 phit[2 x + he] - 0.15 phit[2 x - he];

Plot[{phiu[x], phiun[x]}, {x, -6, 6}]
```

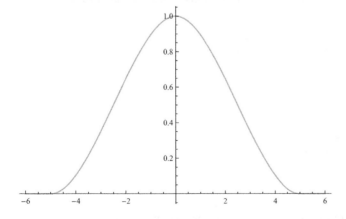

```
Plot[phit[x], {x, -he, he}];

dphit[x_] = D[phit[x], x];

phit[2.5]

0.625

phit[-2.5]

-0.625

dphit[2.5]

-0.25

dphit[-2.5]

-0.25

phitn[x_] = 0.5 phit[2 x] + 0.625 phiu[2 x - he] -
    0.625 phiu[2 x + he] - 0.125 phit[2 x + he] - 0.125 phit[2 x - he];

Plot[{phit[x], phitn[x], 0.5 phit[2 x], 0.625 phiu[2 x - he],
  -0.625 phiu[2 x + he], -0.125 phit[2 x + he], -0.125 phit[2 x - he]}, {x, -he, he}]
```

Mathematica program 6.8 shows the use of a lifting scheme for generation of a Hermite wavelet.

Program 6.8:

```
(* Use of lifting scheme to generate wavelets for Hermite scaling functions *)
he = 5;
```

$$phiu[x\_] = \begin{cases} 1 - 3 (x / he)^2 + 2 (x / he)^3 & 0 \le x \le he \\ 1 - 3 (x / he)^2 - 2 (x / he)^3 & -he \le x \le 0 \end{cases};$$

```
Plot[{phiu[x]}, {x, -6, 6}];

Plot[{phiu[2 x]}, {x, -6, 6}];
```

$$phit[x\_] = \begin{cases} x (1 - x / he)^2 & 0 \le x \le he \\ x (1 + x / he)^2 & -he \le x \le 0 \end{cases};$$

```
Plot[phit[x], {x, -he, he}];

Plot[{phiu[2 x], phiu[x - he / 2], phiu[x + he / 2], phit[x - he / 2], phit[x + he / 2]},
  {x, -2 he, 2 he}, PlotRange → All]
```

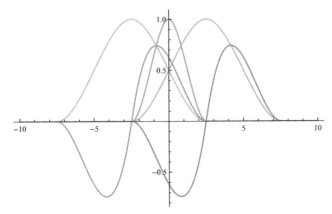

```
psiu[x_] = phiu[2 x] -
    (s1 * phiu[x - he / 2] + s2 * phiu[x + he / 2] + s3 * phit[x - he / 2] + s4 * phit[x + he / 2]);
```

$$\int_{-10}^{10} psiu[x]\, dx$$

$$-\frac{5}{2}\,(-1 + 2\, s1 + 2\, s2)$$

$$\int_{-10}^{10} x * psiu[x]\, dx$$

$$-\frac{25}{6}\,(3\, s1 - 3\, s2 + 2\, s3 + 2\, s4)$$

$$\int_{-10}^{10} x\verb|^|2 * psiu[x]\, dx$$

$$-\frac{25}{12}\,(-1 + 23\, s1 + 23\, s2 + 20\, s3 - 20\, s4)$$

$$\int_{-10}^{10} x\verb|^|3 * psiu[x]\, dx$$

$$-\frac{125}{168}\,(273\, s1 - 273\, s2 + 290\, s3 + 290\, s4)$$

```
LinearSolve[
  {{2, 2, 0, 0}, {3, -3, 2, 2}, {23, 23, 20, -20}, {273, -273, 290, 290}}, {1, 0, 1, 0}]
```

$$\left\{\frac{1}{4},\ \frac{1}{4},\ -\frac{21}{80},\ \frac{21}{80}\right\}$$

```
psiu[x_] = phiu[2 x] + (-1/4) * phiu[x - he / 2] +
    (-1/4) * phiu[x + he / 2] + (21/80) * phit[x - he / 2] + (-21/80) * phit[x + he / 2];

Plot[{psiu[x], phiu[2 x], (-1/4) * phiu[x - he / 2], (-1/4) * phiu[x + he / 2],
    (21/80) * phit[x - he / 2], (-21/80) * phit[x + he / 2]}, {x, -2 he, 2 he}, PlotRange → All]
```

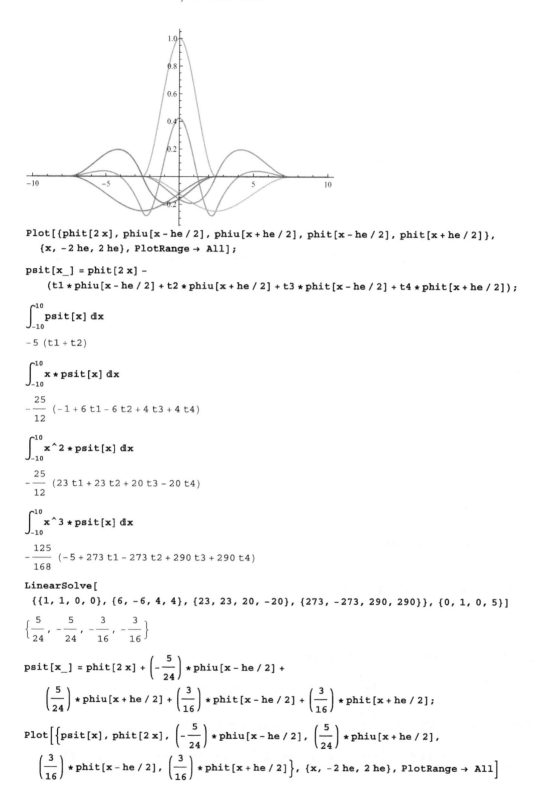

```
Plot[{phit[2 x], phiu[x - he / 2], phiu[x + he / 2], phit[x - he / 2], phit[x + he / 2]},
  {x, -2 he, 2 he}, PlotRange → All];
```

```
psit[x_] = phit[2 x] -
  (t1 * phiu[x - he / 2] + t2 * phiu[x + he / 2] + t3 * phit[x - he / 2] + t4 * phit[x + he / 2]);
```

$$\int_{-10}^{10} psit[x] \, dx$$

-5 (t1 + t2)

$$\int_{-10}^{10} x * psit[x] \, dx$$

$-\dfrac{25}{12}$ (-1 + 6 t1 - 6 t2 + 4 t3 + 4 t4)

$$\int_{-10}^{10} x^2 * psit[x] \, dx$$

$-\dfrac{25}{12}$ (23 t1 + 23 t2 + 20 t3 - 20 t4)

$$\int_{-10}^{10} x^3 * psit[x] \, dx$$

$-\dfrac{125}{168}$ (-5 + 273 t1 - 273 t2 + 290 t3 + 290 t4)

```
LinearSolve[
  {{1, 1, 0, 0}, {6, -6, 4, 4}, {23, 23, 20, -20}, {273, -273, 290, 290}}, {0, 1, 0, 5}]
```

$\left\{ \dfrac{5}{24}, -\dfrac{5}{24}, -\dfrac{3}{16}, -\dfrac{3}{16} \right\}$

```
psit[x_] = phit[2 x] + (-5/24) * phiu[x - he / 2] +
  (5/24) * phiu[x + he / 2] + (3/16) * phit[x - he / 2] + (3/16) * phit[x + he / 2];
```

$$psit[x\_] = phit[2 x] + \left(-\frac{5}{24}\right) * phiu[x - he / 2] +$$
$$\left(\frac{5}{24}\right) * phiu[x + he / 2] + \left(\frac{3}{16}\right) * phit[x - he / 2] + \left(\frac{3}{16}\right) * phit[x + he / 2];$$

$$Plot\left[\left\{psit[x], phit[2 x], \left(-\frac{5}{24}\right) * phiu[x - he / 2], \left(\frac{5}{24}\right) * phiu[x + he / 2],\right.\right.$$
$$\left.\left.\left(\frac{3}{16}\right) * phit[x - he / 2], \left(\frac{3}{16}\right) * phit[x + he / 2]\right\}, \{x, -2 he, 2 he\}, PlotRange → All\right]$$

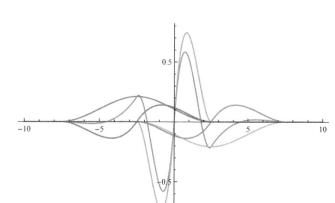

Example 6.5:

As discussed in equation 6.20, it is possible to decouple the refinement process in some cases with the help of scale orthogonal wavelets [8]. The example shows the decoupling process in the following equation

$$-\frac{d^2u}{dx^2}+u+x^2=0 \qquad \text{for} \quad 0<x<L \tag{6.24a}$$

and the boundary conditions

$$u(0)=0, \qquad u(1)=0 \tag{6.24b}$$

The bilinear functional can be defined as

$$a(u,v)=\int_\Omega\left(\frac{du}{dx}\frac{dv}{dx}+uv\right)dx \tag{6.25}$$

We can use linear shape functions or hat basis functions as scaling functions. We use a hierarchical basis function as a lazy wavelet modified with the help of the lifting scheme

$$\psi_k^j=\phi_{2k-1}^{j+1}-\sum_i S_i\phi_i^j$$

The coupling term of the stiffness matrix will be

$$a\left(\psi_k^j,\phi_l^j\right)=a\left(\phi_{2k-1}^{j+1},\phi_l^j\right)-a\left(\sum_i S_i\phi_i^j,\phi_l^j\right) \tag{6.26}$$

for decoupling

$$a\left(\psi_k^j,\ \phi_l^j\right)=0$$

For scale-orthogonality, we can get the lifting coefficients using the following equation

$$a\left(\sum_i S_i \phi_i^j, \phi_l^j\right) = a\left(\phi_{2k-1}^{j+1}, \phi_l^j\right) \tag{6.27}$$

Mathematica Program 6.9:

```
(* Mathematica Program for scale orthogonal basis functions *)
phi01[x_] = Piecewise[{{4x, 0 ≤ x ≤ 1/4}, {(2-4x), 1/4 ≤ x ≤ 1/2}}];
Plot[phi01[x], {x, 0, 1}];
phi02[x_] = Piecewise[{{(4x-1), 1/4 ≤ x ≤ 1/2}, {(3-4x), 1/2 ≤ x ≤ 3/4}}];
Plot[phi02[x], {x, 0, 1}];
phi03[x_] = Piecewise[{{(4x-2), 1/2 ≤ x ≤ 3/4}, {(4-4x), 3/4 ≤ x ≤ 1}}];
Plot[phi03[x], {x, 0, 1}];
w01[x_] = Piecewise[{{8x, 0 ≤ x ≤ 1/8}, {(2-8x), 1/8 ≤ x ≤ 1/4}}];
Plot[w01[x], {x, 0, 1}];
w02[x_] = Piecewise[{{(8x-2), 1/4 ≤ x ≤ 3/8}, {(4-8x), 3/8 ≤ x ≤ 1/2}}];
Plot[w02[x], {x, 0, 1}];
w03[x_] = Piecewise[{{(8x-4), 1/2 ≤ x ≤ 5/8}, {(6-8x), 5/8 ≤ x ≤ 3/4}}];
Plot[w03[x], {x, 0, 1}];
w04[x_] = Piecewise[{{(8x-6), 3/4 ≤ x ≤ 7/8}, {(8-8x), 7/8 ≤ x ≤ 1}}];
Plot[w04[x], {x, 0, 1}, PlotRange → All];
(bigw0[x_] = {{w01[x]}, {w02[x]}, {w03[x]}, {w04[x]}}) // MatrixForm;
(bigphi0[x_] = {{phi01[x]}, {phi02[x]}, {phi03[x]}}) // MatrixForm;
(c0 = Integrate[∂ₓbigphi0[x].Transpose[∂ₓbigphi0[x]] -
    bigphi0[x].Transpose[bigphi0[x]], {x, 0, 1}]) // MatrixForm;
(b01 = Integrate[
    ∂ₓbigphi0[x].Transpose[∂ₓbigw0[x]] - bigphi0[x].Transpose[bigw0[x]], {x, 0, 1}]);
(s01 = Inverse[c0].b01) // MatrixForm;
(bignw0[x_] = bigw0[x] - Transpose[s01].bigphi0[x]) // MatrixForm;
(sum = Partition[Flatten[{{bigphi0[x]}, {bignw0[x]}}], 1]) // MatrixForm;
(K0 = Integrate[∂ₓbigphi0[x].Transpose[∂ₓbigphi0[x]] -
    bigphi0[x].Transpose[bigphi0[x]], {x, 0, 1}]) // MatrixForm;
(C0 = Integrate[∂ₓbignw0[x].Transpose[∂ₓbignw0[x]] -
    bignw0[x].Transpose[bignw0[x]], {x, 0, 1}]) // MatrixForm;
(K = Integrate[∂ₓsum.Transpose[∂ₓsum] - sum.Transpose[sum], {x, 0, 1}]) // MatrixForm;
(F0 = Integrate[-x^2 * bigphi0[x], {x, 0, 1}]) // MatrixForm;
(G0 = Integrate[-x^2 * bignw0[x], {x, 0, 1}]) // MatrixForm;
```

```
(u0 = Inverse[K0].F0) // MatrixForm;

(r0 = Inverse[C0].G0) // MatrixForm;

phi[x_] = Transpose[u0].bigphi0[x] + Transpose[r0].bignw0[x];

phi[.1875]
```
{{-0.0176219}}

```
Plot[phi[x], {x, 0, 1}]
```

```
w11[x_] = Piecewise[{{16 x, 0 ≤ x ≤ 1/16}, {2 - 16 x, 1/16 ≤ x ≤ 2/16}}];

Plot[w11[x], {x, 0, 1}, PlotRange → All];

w12[x_] = Piecewise[{{16 x - 2, 2/16 ≤ x ≤ 3/16}, {4 - 16 x, 3/16 ≤ x ≤ 4/16}}];

Plot[w12[x], {x, 0, 1}, PlotRange → All];

w13[x_] = Piecewise[{{16 x - 4, 4/16 ≤ x ≤ 5/16}, {6 - 16 x, 5/16 ≤ x ≤ 6/16}}];
w14[x_] = Piecewise[{{16 x - 6, 6/16 ≤ x ≤ 7/16}, {8 - 16 x, 7/16 ≤ x ≤ 8/16}}];
w15[x_] = Piecewise[{{16 x - 8, 8/16 ≤ x ≤ 9/16}, {10 - 16 x, 9/16 ≤ x ≤ 10/16}}];

Plot[w13[x], {x, 0, 1}, PlotRange → All];

Plot[w14[x], {x, 0, 1}, PlotRange → All];

Plot[w15[x], {x, 0, 1}, PlotRange → All];

w16[x_] = Piecewise[{{16 x - 10, 10/16 ≤ x ≤ 11/16}, {12 - 16 x, 11/16 ≤ x ≤ 12/16}}];

Plot[w16[x], {x, 0, 1}, PlotRange → All];

w17[x_] = Piecewise[{{16 x - 12, 12/16 ≤ x ≤ 13/16}, {14 - 16 x, 13/16 ≤ x ≤ 14/16}}];

Plot[w17[x], {x, 0, 1}, PlotRange → All];

w18[x_] = Piecewise[{{16 x - 14, 14/16 ≤ x ≤ 15/16}, {16 - 16 x, 15/16 ≤ x ≤ 1}}];

Plot[w18[x], {x, 0, 1}, PlotRange → All];

(bigw1[x_] = {{w11[x]}, {w12[x]}, {w13[x]},
     {w14[x]}, {w15[x]}, {w16[x]}, {w17[x]}, {w18[x]}}) // MatrixForm;

(c1 = Integrate[∂_x bignw0[x].Transpose[∂_x bignw0[x]] -
     bignw0[x].Transpose[bignw0[x]], {x, 0, 1}]) // MatrixForm;
```

```
(b02 = Integrate[∂ₓbigphi0[x].Transpose[∂ₓbigw1[x]] –
    bigphi0[x].Transpose[bigw1[x]], {x, 0, 1}]) // MatrixForm;

(b12 = Integrate[∂ₓbignw0[x].Transpose[∂ₓbigw1[x]] –
    bignw0[x].Transpose[bigw1[x]], {x, 0, 1}]) // MatrixForm;

(s02 = Inverse[c0].b02) // MatrixForm;

(s12 = Inverse[c1].b12) // MatrixForm;

(bignw1[x_] = bigw1[x] – Transpose[s02].bigphi0[x] – Transpose[s12].bignw0[x]) //
  MatrixForm;

(sum1 = Partition[Flatten[{{bigphi0[x]}, {bignw0[x]}, {bignw1[x]}}], 1]) // MatrixForm;

(K = Integrate[∂ₓsum1.Transpose[∂ₓsum1] – sum1.Transpose[sum1], {x, 0, 1}]) // MatrixForm;

(K0 = Integrate[∂ₓbigphi0[x].Transpose[∂ₓbigphi0[x]] –
    bigphi0[x].Transpose[bigphi0[x]], {x, 0, 1}]) // MatrixForm;

(C0 = Integrate[∂ₓbignw0[x].Transpose[∂ₓbignw0[x]] –
    bignw0[x].Transpose[bignw0[x]], {x, 0, 1}]) // MatrixForm;

(C1 = Integrate[∂ₓbignw1[x].Transpose[∂ₓbignw1[x]] –
    bignw1[x].Transpose[bignw1[x]], {x, 0, 1}]) // MatrixForm;

(F0 = Integrate[-x^2 * bigphi0[x], {x, 0, 1}]) // MatrixForm;

(G0 = Integrate[-x^2 * bignw0[x], {x, 0, 1}]) // MatrixForm;

(G1 = Integrate[-x^2 * bignw1[x], {x, 0, 1}]) // MatrixForm;

(u0 = Inverse[K0].F0) // MatrixForm;

(r0 = Inverse[C0].G0) // MatrixForm;

(r1 = Inverse[C1].G1) // MatrixForm;

phil[x_] = Transpose[u0].bigphi0[x] + Transpose[r0].bignw0[x] + Transpose[r1].bignw1[x];

Plot[phil[x], {x, 0, 1}]
```

6.7 NON-STANDARD FORM OF MATRIX AND ITS SOLUTION

The multi-resolution analysis of Non-Standard forms (NS-forms) is used to develop fast direct multi-resolution methods by LU factorization for solving systems of linear algebraic equations arising in elliptic problems. The factorization of the NS-form is different to the standard LU factorization in several ways. Gines et al. [13] have shown that for finite accuracy ϵ, the factorization of NS-forms requires $O(N(-log\epsilon)^2)$ operations, and forward and back substitutions requires $O(N\ (-log\epsilon))$ procedures.

Using wavelet basis function, differential equations or integro-differential are transformed to algebraic equations and can be expressed in matrix form as

$$[K]\{U\}=\{F\}$$

Instead of the Beylkin [14] approach of generating NS-form by using wavelet basis function, we transformed finite element stiffness matrix in NS-form.

$$[w]^T[K][w]\begin{bmatrix} d_1 \\ d_2 \\ \vdots \\ d_m \\ s_1 \\ s_2 \\ \vdots \\ s_n \end{bmatrix}=[w]^T\begin{bmatrix} f_1 \\ f_2 \\ \vdots \\ f_n \end{bmatrix} \tag{6.28}$$

We can use transformation matrix [2] as,

$$[w]=\begin{bmatrix} Q & | & P \end{bmatrix}$$

After one level of transformation, the partitioned matrix can be represented as

$$\begin{bmatrix} [A_1] & [B_1] \\ [C_1] & [K_1] \end{bmatrix}\begin{Bmatrix} \{d_1\} \\ \{s_1\} \end{Bmatrix}=\begin{Bmatrix} \{g_1\} \\ \{f_1\} \end{Bmatrix}$$

Another transformation will generate a matrix which can be shown as

$$\begin{bmatrix} [A_1] & [B_1] & \\ [C_1] & [A_2] & [B_2] \\ & [C_2] & [K_2] \end{bmatrix}\begin{Bmatrix} \{d_1\} \\ \{d_2\} \\ \{s_2\} \end{Bmatrix}=\begin{Bmatrix} \{g_1\} \\ \{g_2\} \\ \{f_2\} \end{Bmatrix}$$

This standard form of matrix can be solved by usual process like LU decomposition. The non-standard operator can be expressed as

$$\begin{bmatrix} [A_1] & [B_1] & & \\ [C_1] & & & \\ & & [A_2] & [B_2] \\ & & [C_2] & [K_2] \end{bmatrix}\begin{Bmatrix} \{d_1\} \\ \{s_1\} \\ \{d_2\} \\ \{s_2\} \end{Bmatrix}=\begin{Bmatrix} \{g_1\} \\ \{f_1-h_1\} \\ \{gh_1\} \\ \{fh_1\} \end{Bmatrix}$$

To solve it, LU type decomposition is used, but the multiplication of upper and lower triangular matrices are not same as the ordinary upper and lower triangular matrices. The LU decomposition of NS-operator can be expressed as

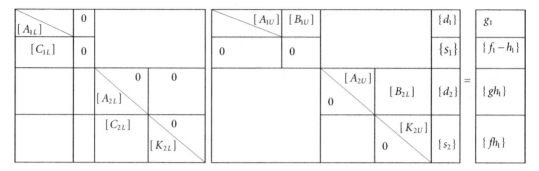

We apply ordinary LU decomposition to A_1

$$[A_{1L}][A_{1U}] = [A_1] \tag{6.29}$$

Also we have,

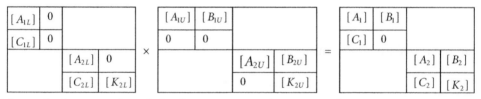

The multiplication of two left side matrices can be expressed as

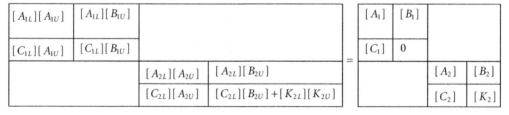

We can get the upper and lower part of the matrix as

$$[B_{1U}] = [A_{1L}]^{-1}.[B_1] \tag{6.30a}$$

$$[C_{1L}] = [C_1].[A_{1U}]^{-1}. \tag{6.30b}$$

Since

$$[C_{1L}][B_{1U}] \neq 0.$$

Therefore

$$[A_{2L}][A_{2U}] \neq [A_2]$$

$$[A_{2L}][B_{2U}] \neq [B_2]$$

$$[C_{2L}][A_{2U}] \neq [C_2]$$

$$[C_{2L}][B_{2U}] + [K_{2L}][K_{2U}] \neq [K_2]$$

This requires the modification at lower level by transformation

$$[Q \mid P]^T.[[C_{1L}][B_{1U}]].[Q \mid P] = \begin{bmatrix} [\bar{A}_2] & [\bar{B}_2] \\ [\bar{C}_2] & [\bar{K}_2] \end{bmatrix}, \tag{6.31}$$

and modify the relations as

$$[A_2] = [A_{2L}][A_{2U}] + [\bar{A}_2] \tag{6.32a}$$

$$[B_2] = [A_{2L}][B_{2U}] + [\bar{B}_2] \tag{6.32b}$$

$$[C_2] = [C_{2L}][A_{2U}] + [\bar{C}_2] \tag{6.32c}$$

$$[K_2] = [C_{2L}][B_{2U}] + [K_{2L}][K_{2U}] + [\bar{K}_2] \tag{6.32d}$$

We apply ordinary LU decomposition to get $[A_{2L}]$ and $[A_{2U}]$

$$[A_{2L}][A_{2U}] = [A_2] - [\bar{A}_2], \tag{6.33}$$

and find the other matrices

$$[B_{2U}] = [A_{2L}]^{-1}.([B_2] - [\bar{B}_2]) \tag{6.34a}$$

$$[C_{2L}] = ([C_2] - [\bar{C}_2]).[A_{2U}]^{-1} \tag{6.34b}$$

We again use ordinary LU decomposition to get $[K_{2L}]$ and $[K_{2U}]$

$$[K_{2L}][K_{2U}] = [K_2] - [\bar{K}_2] - [C_{2L}][B_{2U}]. \tag{6.35}$$

Now the equation can be solved in two steps: *forward sweep* and *backward sweep*.

Forward sweep,

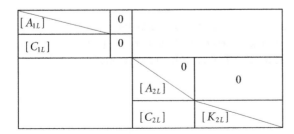 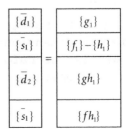

Here, $\bar{d}_j$ and $\bar{s}_j$ are found by the following equations,

$$\{\bar{d}_1\} = [A_{1L}]^{-1}.\{g_1\} \tag{6.36a}$$

$$\{h_1\} = \{f_1\} - [C_{1L}].\{\bar{d}_1\} \tag{6.36b}$$

$$[\{gh_1\} \mid \{fh_1\}]^T = [Q \mid P]^T\{h_1\} \tag{6.37}$$

$$\{\bar{d}_2\} = [A_{2L}]^{-1}.\{gh_1\} \tag{6.38a}$$

$$\{\bar{s}_2\} = [K_{2L}]^{-1}.(\{fh_1\} - [C_{2L}].\{\bar{d}_2\}) \tag{6.38b}$$

Backward sweep,

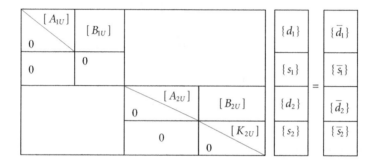

Here, $\{d_j\}$ and $\{s_j\}$ are found by the following equations,

$$\{s_2\} = [K_{2U}]^{-1}.\{\bar{s}_2\} \tag{6.39a}$$

$$\{d_2\} = [A_{2U}]^{-1}.(\{\bar{d}_2\} - [B_{2U}].\{s_2\}) \tag{6.39b}$$

$$\{s_1\} = [Q \mid P].[\{d_2\} \mid \{s_2\}]^T \tag{6.40}$$

$$\{d_1\} = [A_{1U}]^{-1}.(\{\bar{d}_1\} - [B_{1U}].\{s_1\}) \tag{6.41}$$

The MATLAB program 6.10 presents the basic concept of the NS operator. It can be improved with the help of algorithm of Gines et al. [13].

Program 6.10:

```
% Program 6.10: Non-standard Wavelet Transform
% Static one dimensional finite element matrix [aa] .
% Haar wavelet transform is used for standard and
non-standard operator.
clc
clear all
```

```
aa=[ 2 -1  0  0  0  0  0  0
    -1  2 -1  0  0  0  0  0
     0 -1  2 -1  0  0  0  0
     0  0 -1  2 -1  0  0  0
     0  0  0 -1  2 -1  0  0
     0  0  0  0 -1  2 -1  0
     0  0  0  0  0 -1  2 -1
     0  0  0  0  0  0 -1  1];
pqhaar8=[-0.7071       0        0       0     0.7071     0       0      0
          0.7071       0        0       0     0.7071     0       0      0
             0     -0.7071      0       0        0     0.7071    0      0
             0      0.7071      0       0        0     0.7071    0      0
             0       0       -0.7071    0        0       0    0.7071    0
             0       0        0.7071    0        0       0    0.7071    0
             0       0          0    -0.7071  0          0       0   0.7071
             0       0          0     0.7071  0          0       0   0.7071];

pqhaar4=[-0.7071          0         0.7071          0
          0.7071          0         0.7071          0
             0        -0.7071          0         0.7071
             0         0.7071          0         0.7071];

a0=zeros(4,4);
a00=zeros(4,2);
aI=[1 0 0 0
    0 1 0 0
    0 0 1 0
    0 0 0 1];

pq4=[aI          a0
     a0          pqhaar4 ];

a8=pqhaar8'*aa*pqhaar8;
astand=pq4'*a8*pq4;

load1=[0 0 0 0 1 1 1 1]';
wload1=pqhaar8'*load1;
wload2=pq4'*wload1;
tic
 original=inv(aa)*load1
toc
%standard wavelet transform
tic
result4=inv(astand)*wload2;
result0=pq4*result4;
sresult=pqhaar8*result0
toc
```

```
%defining matrices
K1=a8(5:8,5:8);
A1=a8(1:4,1:4);
B1=a8(1:4,5:8);
C1=a8(5:8,1:4);

a4=pqhaar4'*K1*pqhaar4;

A2=a4(1:2,1:2);
B2=a4(1:2,3:4);
C2=a4(3:4,1:2);
K2=a4(3:4,3:4);

        tic
        [A1L,A1U]=lu(A1);
        B1U=inv(A1L)*B1;
        C1L=C1*inv(A1U);

        K1bar=pqhaar4'*(C1L*B1U)*pqhaar4;
        A2bar=K1bar(1:2,1:2);
        B2bar=K1bar(1:2,3:4);
        C2bar=K1bar(3:4,1:2);
        K2bar=K1bar(3:4,3:4);

        [A2L,A2U]=lu(A2-A2bar);
        B2U=inv(A2L)*(B2-B2bar);
        C2L=(C2-C2bar)*inv(A2U);

        [K2L,K2U]=lu(K2-K2bar-C2L*B2U);
% forward sweep
f1=wload1(5:8);
g1=wload1(1:4);

d1bar=inv(A1L)*g1;
 h1bar=f1-C1L*d1bar;
gfh1=pqhaar4'*h1bar;
gh1=gfh1(1:2);
fh1=gfh1(3:4);
d2bar=inv(A2L)*(gh1);
s2bar=inv(K2L)*(fh1-C2L*d2bar);
% backward sweep
s2=inv(K2U)*s2bar;
d2=inv(A2U)*(d2bar-B2U*s2);
ss(1:2)=d2;
ss(3:4)=s2;
s1=pqhaar4*ss';
d1=inv(A1U)*(d1bar-B1U*s1);
```

```
nsresult2=[d1
         d2
         s2];
     nsresult1=pq4*nsresult2;
     nsresult=pqhaar8*nsresult1

toc
```

6.8 MULTIGRID METHOD

Jacobi, Gauss–Seidel and other relaxation iterative solvers are used to solve linear simultaneous algebraic equations involving large number of unknowns

$$[K]\{u\}=\{f\},$$

The classical iterative solvers quickly remove high frequency error but the low frequency error slows down the convergence. The multigrid approach is often used to reduce the number of iterations. It is called multigrid because the method uses both coarse and fine grid to reduce the number of iterations. Figure 6.7 shows an example of fine and coarse grid. The coarse grid is effective in eliminating the components of low frequency error whereas the high frequency components are damped by fine grid. Wavelets can be used for switching progressively from fine to coarse and back to fine grids and build the iterative solution. We refer h for fine grid and H for coarse grid, and use the following steps for the two-grid method:

1. Transform matrix for course grid $[K^H]=[P]^T[K][P]$

2. Assume initial approximate solution $\{u^h\}=\{0\}$

3. Apply a few steps of Gauss–Seidel or Jacobi to $[K]\{u\}=\{f\}$ to update $\{u^h\}$

4. Compute the residual for this solution $\{r^h\}=\{f\}-[K]\{u^h\}$

5. Transform the residual to the coarse $\{r^H\}=[P]\{r^h\}$

6. Solve the error equation at coarse $[K^H]\{e^H\}=\{r^H\}$

7. Update the approximation $\{u^h\}=\{u^h\}+[P]^T\{e^H\}$

8. Repeat the same from step 3 till the desired accuracy is achieved.

The MATLAB program shows some improvement due to the multigrid approach. Readers can find many good works and improve their programs [15–18].

```
% Program 6.11: Wavelet based multigrid method
% Static one dimensional finite element matrix [aa] is used.
% Haar wavelet is used.
clc
clear all
K=[ 2 -1  0  0  0  0  0  0
   -1  2 -1  0  0  0  0  0
    0 -1  2 -1  0  0  0  0
    0  0 -1  2 -1  0  0  0
    0  0  0 -1  2 -1  0  0
    0  0  0  0 -1  2 -1  0
    0  0  0  0  0 -1  2 -1
    0  0  0  0  0  0 -1  1];
f=[0 0 0 0 1 1 1 1]';

tic
original=inv(K)*f % for the comparison with direct solver
toc

Phaar8 = [  0.7071     0        0        0
            0.7071     0        0        0
               0     0.7071     0        0
               0     0.7071     0        0
               0        0     0.7071     0
               0        0     0.7071     0
               0        0        0     0.7071
               0        0        0     0.7071];

    Uh=zeros(1,8)';
    KH=Phaar8'*K*Phaar8;
    eH= zeros(1,4)';

    omega=0.5;
    iter=0;
% iterative solver
    tic
    for j=1:300
    for i=1:2
        Uold=Uh;
        Uh=Uold+omega*(f-K*Uold);
        err=100*(norm(Uold-Uh)/norm(Uh));
        iter=iter+1;
    end

% Loop for Multigrid
% To remove multigrid, comment the following multigrid part.
        rh=f-K*Uh;
        rH=Phaar8'*rh;
```

```
        for k=1:2
        eHold=eH;
        eH=eHold+omega*(rH-KH*eHold);
        iter=iter+1;
        end
        Uh=Uh+Phaar8*eH;
% Multigrid ends here
            if(err<0.005)
               break;
            end

    end

    iter    % to test iteration
    Uh      % iterative solution
    toc
```

REFERENCES

1. P.S. Addison (2002), *The Illustrated Wavelet Transform Handbook*, Institute of Physics Publishing, Bristol and Philadelphia, PA.
2. E.J. Stollnitz, T.D. Derose and D.H. Salesin (1996), *Wavelets for Computer Graphics, Theory and Applications*, Morgan Kaufmann Publishers, San Francisco, CA.
3. G. Strang and T. Nguyen (1996), *Wavelets and Filter Banks*, Wellesley-Cambridge Press, Stockport, UK.
4. A. Jensen and A.la Cour-Harbo (2009), *Ripples in Mathematics*, Springer-Verlag, New Delhi, India.
5. W. Sweldens (1998), "The lifting scheme: A construction of second generation wavelets," *SIAM Journal of Mathematical Analysis*, Vol. 29, No. 2, pp. 511–546.
6. I. Daubechies and W. Sweldens (1998), "Factoring wavelet transform into lifting steps," *The Journal of Fourier Analysis and Applications*, Vol. 4, No. 3, pp. 247–269.
7. S.F. D'Heedene, *An Operator-Customized Wavelet Finite Element Approach for the Adaptive Solution of Second-order Partial Differential Equations on Unstructured Meshes*, Ph. D. Thesis, MIT, February 2005.
8. R. Sudarshan, *Operator-Adapted Finite Element Wavelets: Theory and Applications to A Posteriori Error Estimation and Adaptive Computational Modeling*, Ph. D. Thesis, MIT, June 2005.
9. K. Amaratunga and R. Sudarshan (2006), "Multiresolution modeling with operator-customized wavelet derived from finite elements," *Computer Methods in Applied Mechanics and Engineering*, Vol. 195, pp. 2509–2532.
10. S.M. Quraishi and K. Sandeep (2011), "A second generation wavelet based finite elements on triangulations," *Computational Mechanics*, Vol. 48, No.2, pp. 163–174.
11. R. Sudarshan, S. Heedene and K. Amaratunga (2003), "A multiresolution finite element method using second generation Hermitemultiwavelets," *Second MIT Conference on Computational Fluid and Solid Mechanics*, pp. 1–6.
12. J.N. Reddy (2005), *An Introduction to the Finite Element Method* (3rd ed.), McGraw-Hill Education, New York.
13. D. Gines, G. Beylkin and J. Dunn (1998), "LU Factorization of non-standard forms and direct multiresolution solvers," *Applied and Computational Harmonic Analysis*, Vol. 5, pp. 156–201.

14. G. Beylkin (1992), "On the representation of operators in bases of compactly supported wavelets," *SIAM Journal of Numerical Analysis*, Vol. 6, pp. 1716–1740.

15. S. Goedecker and C. Chauvin (2003), "Combining multigrid and wavelet idea to construct more efficient multiscale algorithm for the solution of Poisson's equation," *Journal of Theoretical and Computational Chemistry*, Vol. 2, No. 2, pp. 483–495.

16. A. Avudainayagam and C. Vani (2004), "Wavelet based multigrid methods for linear and nonlinear elliptic partial differential equations," *Applied Mathematics and Computations*, Vol. 148, pp. 307–320.

17. N.M. Bujurke, C.S. Salimath, R.B. Kudenatti and S.C. Shiralashetti (2007), "A fast wavelet multigrid method to solve elliptic partial differential equations," *Applied Mathematics and Computations* Vol. 185, pp. 667–680.

18. B. Engquist and E. Luo (1996), "The multigrid method based on a wavelet transformation and Schur complement," *Research Report*, Carnegie Mellon University, Pittsburgh, PA.

CHAPTER 7

Error Estimation

A FTER OBTAINING A FINITE element solution, we often want to estimate the error in the solution known as *a-posteriori* error estimation. Several techniques have been developed to provide estimates and bounds for the solution error in a specified norm or a quantity of interest. The effectiveness of some *a-posteriori* error estimators has resulted in efficient algorithms of the adaptive finite element method where either the mesh is locally refined (*h*-method) or a higher order polynomial is used (*p*-method). Such type of adaptive method is necessary for the real-life numerical simulation where the mesh size becomes too large for computation. Various types of *a-posteriori* error estimation techniques are presented in this chapter.

7.1 INTRODUCTION

Most of the partial differential equations characterizing the continuum models are solved by some numerical method. These numerical simulations have different sources of errors. The errors in the finite element method are due to use of: (i) discretization and approximation of domain using a finite number of elements, (ii) use of simplified representation of the field variable in order to reduce the complexity of computation, (iii) approximate numerical process for computation of stiffness matrix, load vector and finally the solution. To study the error estimation technique in FEM, let us revisit the method.

Consider a linear elliptic problem

$$-\Delta u = f \qquad \text{in } \Omega \tag{7.1a}$$

$$u = 0 \qquad \text{on } \Gamma_D \tag{7.1b}$$

$$\hat{n}.\nabla u = g \qquad \text{on } \Gamma_N \tag{7.1c}$$

where $\hat{n}$ denotes the unit outward normal vector. For simplicity, we consider a two-dimensional domain $\Omega \subset \mathbb{R}^2$ with Lipschitz continuous boundary $\partial\Omega$. The given data $f \in L^2(\Omega)$ and $g \in L^2(\Gamma_N)$. The boundary $\partial\Omega = \Gamma_D \bigcup \Gamma_N$ and $\Gamma_D \bigcap \Gamma_N = \varnothing$. For the existence of a solution in Sobolev space $V = \{v \in H^1(\Omega) : v = 0 \text{ on } \Gamma_D\}$, the weak formulation of the equation (7.1) is: find $u \in V$, such that

$$a(u,v)=l(v) \quad \forall v \in V \tag{7.2a}$$

where,

$$a(u,v)=\int_\Omega \nabla u.\nabla v\, d\Omega \tag{7.2b}$$

$$l(v)=\int_\Omega f.v\, d\Omega + \int_{\Gamma_N} g.v\, ds \tag{7.2c}$$

Equipped with the energy norm

$$\|u\|_E = \sqrt{a(u,u)} \tag{7.3}$$

We are considering only the problem where the unique solution exists. Since it is not possible to consider infinite dimensional space V for numerical simulation, we consider the space V_h.

Let $V_h \subset V$ be a suitable finite element space. The finite element method is to find $u_h \in V_h$ such that

$$a(u_h,v_h)=l(v_h) \quad \forall v_h \in V_h \tag{7.4}$$

Two questions arise when we solve any problem with the help of the finite element method:

1. How close the finite element approximation is to the exact solution.

2. How to ensure continuous improvement of solution with the refinement of mesh.

The first question is difficult to answer because, generally, the exact solution is unknown. The second question, i.e., convergence to the exact solution, is ensured by assuming polynomial with

1. Constant term: This is necessary for the representation of rigid body displacement.

2. Constant strain term: An infinitesimal element will show nearly constant strain behavior; therefore it is necessary to include it.
    This means that we should assume the basis function as

$$u_e^h = a_0 + a_1 x \text{ for one-dimensional}$$

$$u_e^h = a_0 + a_1 x + a_2 y \text{ for two-dimensional}$$

3. If a derivative of order "m" exists in the integral form of the equation, then the polynomial should be at least of the order "m."

These three conditions of interpolation function ensure the convergence of a solution with the refinement of mesh.

The error in the finite element solution is to be calculated to know the answer of the first question. The error in the finite element solution is

$$e_h = u - u_h$$

Since we cannot find the exact solution u in most of the cases, the exact error "e_h" cannot be obtained. Various methods to estimate the error e_h are used. A-priori error estimation provides some information on the total error on the basis of element size and order of polynomial of the domain. On the other hand, a-posteriori error estimation technique identifies the elements with large errors after the analysis of a finite element solution. Computing exact error using a-posteriori technique, where the exact solutions are unknown, will be as difficult as finding the exact solution.

Several a-posteriori error estimation methods have been developed to indicate the distribution of discretization error. It is necessary that the method to find the distribution of error in the finite element mesh should be computationally economical. A good error estimator is very useful for adaptive mesh refinement to improve the accuracy of the solution with optimum computational cost. The adaptive algorithm reduces total error either by increasing the number of elements known as h-method or by increasing the order of polynomial known as p-method. A good amount of mathematical theories on the error estimation techniques are developed, but they should be used with care [1,2]. It is beyond the scope of this chapter to present all the theories of a-posteriori error estimation techniques. However, the chapter will present the basic procedures of recovery based, residual based, goal oriented and hierarchical techniques. Although Mathematica programs are not included in this chapter, we suggest students use it. We advise to test any error estimation technique before its implementation.

7.2 A-PRIORI ERROR ESTIMATION

The method is called a-priori because it is used to estimate the error before computing the finite element solution. If the interpolation function is satisfying the three conditions as stated in Section 7.1, then the symmetric bilinear form gives the best approximation in the finite element space V_h, in the energy norm, i.e.,

$$\left\| u - u_h \right\|_{H^1(\Omega)} \le C h^p \left\| u \right\|_{H^{p+1}(\Omega)}, \quad p > 0 \tag{7.5a}$$

and error in L^2-Norm

$$\left\| u - u_h \right\|_{L^2(\Omega)} \le Ch^{p+1} \left\| u \right\|_{H^{p+1}(\Omega)} \tag{7.5b}$$

This can be used to derive the rate of convergence. For example, if u_h is the approximate solution with meshes of size "h" and u_{h2} is the solution for size "$h/2$" then

$$\frac{\left\| u - u_h \right\|_{H^1(\Omega)}}{\left\| u - u_{h2} \right\|_{H^1(\Omega)}} \le 2^p \tag{7.6}$$

Though it shows the convergence rate, the exact solution cannot be obtained due to *round-off* error of the computer and the accuracy limitation of the numerical integration and matrix solution. The total error can be reduced by uniformly increasing the number of elements and order of polynomial in the whole domain. Such uniform refinement is not desirable as it will increase the computational cost significantly. Therefore, various *a-posteriori* error estimation techniques are developed to identify the elements which are the major source of errors.

7.3 RECOVERY BASED METHOD

The finite element method produces the best approximation of field variables at the nodes. In an elasticity problem, we get piecewise continuous estimates for the element strains and stresses, inside the elements. However, the estimate of flux or gradient of the field variable generally jumps across the element boundaries. The exact stress/strain should be continuous for homogeneous domain. The flux should not be discontinuous in the structure as we find in the finite element discretization. The gradient or strain will be discontinuous but the flux or stress will be continuous at the interface of the material if the structure is made of two homogeneous materials with different modulus of elasticity. If σ_i and σ_j are stresses in two elements with a common boundary, then $\sigma_i - \sigma_j$ indicates the error at the boundary. To find the exact error, we have to find the difference between the exact stress and the stress due to finite element solution. In general, we do not know the exact strain or stress values. Finding an alternative of exact flux for the computation of error is the main issue. In the case of a one-dimensional problem, the easiest approach is to assume average of stress/strain at the node as the exact value. In general, we use a proper interpolation function to represent exact stress/strain.

In a domain of interest, we generate a continuous interpolation function with the help of piecewise continuous stress or strain of the element. This continuous estimate of strain and stress is more accurate than the stress and strain of the finite element. To generate a continuous interpolation function, we have to use some values of stress and strain which are the most accurate. There are some locations within an element where the derivatives are most accurate which are known as super-convergent points [3,4]. In some cases, locations coincide with the Gaussian quadrature points so we can use these points for convenience. Let ε_i be the strain at the super-convergent point i (or Gaussian quadrature point) and

n be the number of points in a small domain where a smooth polynomial is to be created. The polynomial of order m at any location i can be expressed as:

$$\varepsilon_i = a_0 + a_1 x_i + a_2 y_i + \ldots + a_k y_i^m \tag{7.7}$$

The n equations can be written as:

$$\begin{Bmatrix} \varepsilon_1 \\ \varepsilon_2 \\ . \\ . \\ \varepsilon_n \end{Bmatrix} = \begin{bmatrix} 1 & x_1 & y_1 & . & y_1^m \\ 1 & x_2 & y_2 & . & y_2^m \\ . & . & . & . & . \\ . & . & . & . & . \\ 1 & x_n & y_n & . & y_n^m \end{bmatrix} \begin{Bmatrix} a_0 \\ a_1 \\ . \\ . \\ a_k \end{Bmatrix} \tag{7.8}$$

which we represent as

$$\{\varepsilon_n\} = \left[P_n^m \right] \{a_k\} \tag{7.9}$$

Here $k \leq n$. We have to find the coefficients of polynomial $\{a_k\}$. It is not a problem for $k = n$, but if the number of unknowns is less than the number of equations, then the least square method is to be used. It can be expressed in the matrix form as:

$$\{a_k\} = \left(\left[P_n^m \right]^T . \left[P_n^m \right] \right)^{-1} . \left[P_n^m \right]^T \{\varepsilon_n\} \tag{7.10}$$

The smooth polynomial over elements where we want to estimate the error is assumed as the exact polynomial of flux for the computation of error. As the method is easy to implement, many commercial finite element software packages incorporate it. This crude approach of comparing the smoothed gradient with non–post-processed gradient provides good estimates for the true error. Numerous minor improvements are also suggested and used in this original method.

7.4 RESIDUAL BASED ERROR ESTIMATORS

The central issue in the residual based method is to use the interior element residual and jump of the gradient across the element boundary for estimation of the error. For the finite element solution u_h and input data f and g, the residual is

$$R_j = f + \Delta u_h \quad \text{in } \Omega_j \tag{7.11}$$

and the jump across the element boundary γ_i is expressed as

$$J_i = \begin{cases} (n.\nabla u_h + n'.\nabla u_h') & \text{if} & \gamma_i \text{ is not at the boundary} \\ g - n.\nabla u_h & \text{if} & \gamma_i \text{ is at the Neumann boundary} \\ 0 & \text{if} & \gamma_i \text{ is at the Dirichlet boundary.} \end{cases} \tag{7.12}$$

The residual and jump can be used to estimate the error in two different ways known as explicit and implicit residual error estimators. For explicit method, R_j and J_i are employed directly to compute the error norm. In the case of implicit error estimators, a set of linear algebraic equations is formed using R_j and J_i and solved to get the error.

The error for the explicit residual method is expressed as [1]

$$\|e_h\|^2 \leq \sum_j \eta_j^2 \tag{7.13}$$

where,

$$\eta_j = c\left[\sum_j h_j^2 \|R_j\|_{L^2(\Omega_j)}^2 + \sum_i h_i \|J_i\|_{L^2(\gamma_i)}^2\right]^{1/2} \tag{7.14}$$

Here, h_j is the diameter of the element j. The unknown coefficient "c" on the right side of the above equation can have two different values for two terms. On inter element boundary, $\gamma_i \not\subset \Gamma$, the jump J_i is equally distributed between the two elements sharing the common boundary.

Although constant "c" is unknown, the method can be used for mesh adaptivity. The relative error is used for next level of refinement. It can be calculated by using

$$e_{rel} = \frac{\sum_j \eta_j^2}{\|u_h\|_E} \tag{7.15}$$

If the relative error exceeds a specified tolerance, then mesh is to be enriched. If the error in each element is almost equal, then the mesh is considered optimal. Elements with large error as compared to other elements are refined to improve the solution.

For implicit residual error estimator, let us recall our differential equations and boundary conditions (7.1) for element j

$$\Delta u + f = 0 \qquad \text{in } \Omega_j \tag{7.16a}$$

$$\hat{n}.\nabla u = \frac{\partial u}{\partial n} \qquad \text{on } \Gamma_j \tag{7.16b}$$

There is some residual when we substitute the approximate finite element solution in the above equations, which can be expressed as

$$\Delta u_h + f = R_j \text{ in } \Omega_j \tag{7.17a}$$

$$\hat{n}.\nabla u_h = \frac{\partial u_h}{\partial n} \qquad \text{on element boundary } \Gamma_j \tag{7.17b}$$

Subtracting the two sets of equations for element j, we get

$$\Delta(u - u_h) + R_j = 0 \qquad \text{in } \Omega_j$$

or

$$\Delta(e_j) + R_j = 0 \qquad \text{in } \Omega_j \qquad (7.18a)$$

and

$$\hat{n}.\nabla(e_j) = \frac{\partial u}{\partial n} - \frac{\partial u_h}{\partial n} \qquad \text{on } \Gamma_j \qquad (7.18b)$$

Since, $\dfrac{\partial u}{\partial n}$ is unknown, the unknown true flux is approximated from the finite element solution as the average flux of the two elements which are sharing the boundary:

$$\frac{\partial u}{\partial n} \approx \frac{\partial \bar{u}_h}{\partial n} = \frac{1}{2} \hat{n} \left(\nabla u_h + \nabla u_h' \right) \qquad (7.19)$$

The weak form of equation (7.18) is expressed as

$$a(\bar{e}_j, v) = \int\limits_{\Omega_j} R_j v \, d\Omega + \int\limits_{\Gamma_j} \left(\frac{\partial \bar{u}_h}{\partial n} - \frac{\partial u_h}{\partial n} \right) v \, d\Gamma \qquad \forall v \in V_j \qquad (7.20)$$

The total error can be estimated with the solution $\bar{e}_j$

$$\|e_h\|^2 \approx \sum_j \|\bar{e}_j\|^2 \qquad (7.21)$$

7.5 GOAL ORIENTED ERROR ESTIMATORS

The goal of many finite element computations is to determine specific output data of the finite element approximation. To find an estimate for the error in the output data about a specific quantity, or to find at least an effective mesh to solve for this quantity accurately, energy norm based error estimators are not useful. The relatively new techniques called goal oriented error estimators are developed, which estimate the error in individual quantities of interest using duality techniques.

Let $Q(u)$ denote such a specific quantity. For example, it may be

- The value of solution at a particular point in Ω : $Q(u) = u(x_0)$

- Average value of the solution in a subdomain of special interest Ω_0 in Ω:

$$Q(u) = \frac{1}{|\Omega_0|} \int\limits_{\Omega_0} u(x) \, d\Omega$$

- The flux flowing out over a portion of the boundary (Γ):

$$Q(u) = \int_{\Gamma_0} n.\nabla u \, d\Gamma$$

Consider the primal problem equation (7.4) as the model problem

$$a(u_h, v_h) = l(v_h) \qquad \forall \, v_h \in V_h$$

where,

$$a(u_h, v_h) = \int_{\Omega} \frac{\partial u_h}{\partial x} \frac{\partial v_h}{\partial x} \, d\Omega$$

$$l(v_h) = \int_{\Omega} f.v_h \, d\Omega + \int_{\Gamma_N} g.v_h \, d\Gamma$$

Let us consider "goal" as the integral $\int_{\Omega_0} u(x) \, d\Omega$. To evaluate the value of such a goal via the finite element solutions u_h, we introduce the corresponding linear functional Q and estimate the following value

$$f_0 = Q(u_h) = \frac{1}{|\Omega_0|} \int_{\Omega_0} u_h(x) \, d\Omega \tag{7.22}$$

The primal space V_h can be used for the dual problem also to find a function z_h such that

$$a(z_h, v_h) = Q(v_h) \qquad \forall \, v_h \in V_h \tag{7.23}$$

where,

$$a(z_h, v_h) = \int_{\Omega} \frac{\partial z_h}{\partial x} \frac{\partial v_h}{\partial x} \, d\Omega$$

$$Q(v_h) = \int_{\Omega} f_0.v_h \, d\Omega$$

The solution of this problem is stated as the dual solution and can be inferred as the generalized Green's function, or the influence function, related to the functional $Q(v_h)$.

We have to find an estimate for the error as

$$Q(e_h) = Q(u) - Q(u_h) \tag{7.24}$$

The upper error bound is expressed as [1,5]

$$\left|Q(e_h)\right| \leq \|u - u_h\|_E \|z - z_h\|_E \tag{7.25}$$

Since, u and z are unknown, we use recovery or residual methods to evaluate $u - u_{hE}$ and $z - z_{hE}$. If we use explicit residual error estimator then

$$\left|Q(e_h)\right| \leq \sum_j \omega_j \eta_j \tag{7.26}$$

$$\eta_j = c \left[\sum_j h_j^2 \|R_j\|_{L^2(\Omega_j)}^2 + \sum_i h_j \|J_i\|_{L^2(\gamma_i)}^2 \right]^{\frac{1}{2}} \tag{7.27a}$$

$$\omega_j = c \left[\sum_j h_j^2 \|(R_j)_z\|_{L^2(\Omega_j)}^2 + \sum_i h_j \|(J_i)_z\|_{L^2(\gamma_i)}^2 \right]^{\frac{1}{2}} \tag{7.27b}$$

where $(R_j)_z$ and $(J_i)_z$ denote the element residual and jumps of the gradient to the adjoint problem. The relative error can be obtained as

$$e_{rel} = \frac{\left|Q(e_h)\right|}{\left|Q(u_h)\right|} \tag{7.28}$$

7.6 HIERARCHICAL AND WAVELET BASED ERROR ESTIMATORS

In contrast to other methods of computing error where the current solution is compared with the exact (or approximated exact) solution, this method compares the current solution with the next refined solution. We can consider the hierarchical basis function as a special case of wavelets. In fact, the hierarchical basis functions are also called "Lazy" wavelets. The hierarchical basis function refines the approximation without changing the initial basis functions and their coefficients.

Let the initial finite element space be $\mathcal{V}^j \subset L^2(\Omega)$, and the next refined space be $\mathcal{V}^{j+1} \subset L^2(\Omega)$, i.e., $\mathcal{V}^j \subset \mathcal{V}^{j+1} \subset L^2(\Omega)$. Similar to wavelets, we can define the space of hierarchical basis for the improvement of solution as $\mathcal{W}^j$ such that $\mathcal{V}^{j+1} = \mathcal{V}^j \oplus \mathcal{W}^j$.

Consider the approximation $u_j \in \mathcal{V}^j$ such that

$$a(u_j, v) = f(v) \tag{7.29}$$

for all $v \in \mathcal{V}^j$. The solution holds the approximation property

$$\|u - u_j\| = \inf_{v \in V_j} \|u - v\|. \tag{7.30}$$

Here $\|u\|^2 = a(u,u)$. We can assume that space $\mathcal{V}^j \subset \mathcal{V}^{j+1} \subset L^2(\Omega)$. With space $\mathcal{V}^{j+1}$, Galerkin approximate solution u_{j+1} will satisfy

$$a(u_{j+1}, v) = f(v) \tag{7.31}$$

for all $v \in \mathcal{V}^{j+1}$. Let us assume that the approximate solution u_{j+1} is closer to u than u_j. This is defined in terms of the saturation assumption

$$\|u - u_{j+1}\| \leq \xi \|u - u_j\| \tag{7.32}$$

where $\xi < 1$. Instead of using $e = u - u_j$, [6] used two-scale error in the multiresolution approach

$$r_j = u_{j+1} - u_j \tag{7.33}$$

where $r_j \in W^j$. It means that the two-level error is the projection of true error onto the wavelet space. It is proved that

$$c_1 \|u - u_j\| \leq \|r_j\| \leq c_2 \|u - u_j\| \tag{7.34}$$

where $c_1, c_2 > 0$ are constants and independent of the multi-resolution level j. The wavelet coefficients w_j are the projection of r_j, hence

$$\|w_{j+1}\| \leq \|r_{j+1}\| \leq \|w_j\| \leq \|r_j\| \leq \|u - u_j\|. \tag{7.35}$$

The two consecutive levels of FEM refinements can be used to calculate two scale error $\|r_{j+1}\|$ or $\|r_j\|$. This method will be expensive to estimate error. Therefore, the FE solution is transformed using discrete wavelets. The norm calculated using wavelet coefficients will show a sudden jump in some areas which will indicate the need for mesh refinement in the zone. The cost of computing $\|w_{j+1}\|$ or $\|w_j\|$ is considerably less than the two scale error. The wavelet coefficients can be calculated using usual discrete wavelet transform. To use first generation wavelets, the finite element nodes should be equally spaced. The second generation wavelets can be used for non-uniform grid. The method can also be used with the goal oriented error estimation technique [7].

7.7 MODEL PROBLEM

To understand how to use *a-posteriori* error estimation techniques presented in this chapter, let us consider the following one-dimensional problem:

$$-\frac{d^2 u}{dx^2} = \frac{x^3}{1000} \qquad 0 \leq x \leq 12$$

subjected to the boundary conditions

$$u = 0 \quad \text{at } x = 0,$$

$$\frac{du}{dx} = 0 \quad \text{at } x = 12.$$

The exact solution of the above model problem is

$$u_{exact} = -\frac{x^5}{20000} + \frac{648x}{125}$$

Figure 7.1 shows the discretization of domain for FEM solution (u_h).
Using linear element, we get the following finite element solution

1. For two elements:

$$\begin{bmatrix} u_1 \\ u_2 \\ u_3 \end{bmatrix} = \begin{bmatrix} 0 \\ 30.7152 \\ 49.7664 \end{bmatrix}$$

2. For three elements:

$$\begin{bmatrix} u_1 \\ u_2 \\ u_3 \\ u_4 \end{bmatrix} = \begin{bmatrix} 0 \\ 20.6848 \\ 39.8336 \\ 49.7664 \end{bmatrix}$$

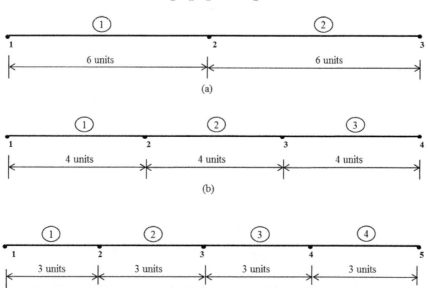

FIGURE 7.1 Discretization of domain (a) two elements, (b) three elements, and (c) four elements.

3. For four elements

$$\begin{bmatrix} u_1 \\ u_2 \\ u_3 \\ u_4 \\ u_5 \end{bmatrix} = \begin{bmatrix} 0 \\ 15.5398 \\ 30.7152 \\ 43.7035 \\ 49.7664 \end{bmatrix}$$

7.7.1 Error by Recovery Based Method

Consider three Gauss points in three elements shown in Figure 7.1b, i.e., $gp_1 = 2.0$, $gp_2 = 6.0$ and $gp_3 = 10.0$.

The strains obtained using the finite element solution at these Gauss points are

$$\frac{du_h}{dx} = 5.1712 \text{ at } gp_1$$

$$\frac{du_h}{dx} = 4.7872 \text{ at } gp_2$$

$$\frac{du_h}{dx} = 2.4832 \text{ at } gp_3$$

We can assume linear interpolation function to approximate the exact strain as $\bar{\varepsilon} = a_0 + a_1 x$. Equating it with the strains obtained using the finite element solution at these Gauss points, we get

$$\text{at } gp_1, \qquad a_0 + 2.a_1 = 5.1712$$

$$\text{at } gp_2 \qquad a_0 + 6.a_1 = 4.7872$$

$$\text{at } gp_2 \qquad a_0 + 10.a_1 = 2.4832$$

We can write it in the matrix form as

$$\left[P_3^1 \right] = \begin{bmatrix} 1 & 2 \\ 1 & 6 \\ 1 & 10 \end{bmatrix}, \{a_2\} = \begin{bmatrix} a_0 \\ a_1 \end{bmatrix}, \{\varepsilon_3\} = \begin{bmatrix} 5.1712 \\ 4.7872 \\ 2.4832 \end{bmatrix}$$

Since the number of equations is more than the number of variables, we use equation (7.10). Solving it, we get

$$\{a_2\} = \begin{bmatrix} 6.1632 \\ -0.336 \end{bmatrix}$$

Therefore, approximation of exact strain is

$$\bar{\varepsilon} = 6.1632 - 0.336x$$

However, the exact strain is

$$\varepsilon = \frac{du_{exact}}{dx} = -\frac{x^4}{4000} + \frac{648}{125}$$

For the same example, let us assume quadratic interpolation function to approximate the exact strain as $\hat{\varepsilon} = a_0 + a_1 x + a_2 x^2$. Equating it with the strains at the Gauss points, we get,

$$\text{at } gp_1, \; a_0 + 2.a_1 + 4.a_2 = 5.1712$$

$$\text{at } gp_2, \; a_0 + 6.a_1 + 36.a_2 = 4.7872$$

$$\text{at } gp_3, \; a_0 + 10.a_1 + 100.a_2 = 2.4832$$

or the matrix form,

$$\begin{bmatrix} 1 & 2 & 4 \\ 1 & 6 & 36 \\ 1 & 10 & 100 \end{bmatrix} \begin{bmatrix} a_0 \\ a_1 \\ a_2 \end{bmatrix} = \begin{bmatrix} 5.1712 \\ 4.7872 \\ 2.4832 \end{bmatrix}$$

We get polynomial coefficients as

$$\begin{bmatrix} a_0 \\ a_1 \\ a_2 \end{bmatrix} = \begin{bmatrix} 4.6432 \\ 0.384 \\ -0.06 \end{bmatrix}$$

Therefore, approximation of exact strain is

$$\hat{\varepsilon} = 4.6432 + 0.384x - 0.06x^2$$

Figure 7.2 shows the exact strain and its linear and quadratic approximations. The approximation of exact strain improves as the number of elements and order of polynomial increases. This approximated exact strain is compared with the strain in the elements to find the elements with large error.

7.7.2 Error by Explicit Residual Estimator

For case (a), after finite element solution, we get the interpolation functions between 1 to 2 and 2 to 3 as

$$u_{1to2} = 5.1192x$$

$$u_{2to3} = 5.1192(12 - x) + 8.2944(-6 + x)$$

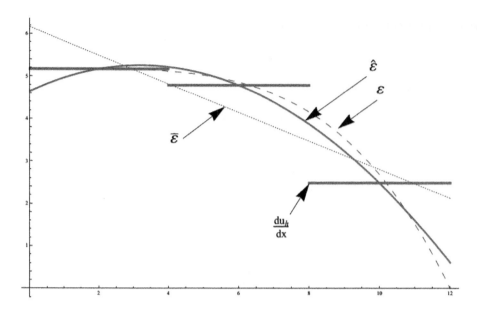

FIGURE 7.2   The exact strain, approximated exact strain and FEM strain.

Since, we assumed linear interpolation functions,

$$\Delta u_{1to2} = \Delta u_{2to3} = 0$$

The residuals $(R_j = f + \Delta u_h)$ for two elements are

$$R_1 = \frac{x^3}{1000} = R_2$$

Therefore,

$$\left\| R_1 \right\|_{L^2(\Omega_1)} = \left[\int_0^6 (R_1)^2 \, dx \right]^{\frac{1}{2}} = 0.199977$$

$$\left\| R_2 \right\|_{L^2(\Omega_2)} = \left[\int_6^{12} (R_2)^2 \, dx \right]^{\frac{1}{2}} = 2.25363$$

Jump at element boundary:

at inter-element boundary, node 2, $J_1 = \nabla u_{1to2} - \nabla u_{2to3} = 1.944$

at Neumann boundary, node 3, $J_2 = 0 - \nabla u_{2to3} = -3.1752$

In this case, element size is $h_1 = h_2 = 6$.

Therefore,

$$\eta_1 = \left[h_1^2 \|R_1\|_{L^2(\Omega_1)}^2 + h_1 \left(\frac{J_1}{2} \right)^2 \right]^{\frac{1}{2}} = 2.66615$$

$$\eta_2 = \left[h_2^2 \|R_2\|_{L^2(\Omega_2)}^2 + h_2 \left\{ \left(\frac{J_1}{2} \right)^2 + (J_2)^2 \right\} \right]^{\frac{1}{2}} = 15.7797$$

and the total error

$$\|e_h\|_E = \eta_1 + \eta_2 = 18.4458$$

The large error at element (2) suggests more elements between node 2 and 3 are required.

7.7.3 Error by Implicit Residual Estimator

Case (a)

$$a(\bar{e}_j, v) = \int_{\Omega_j} R_j v d\Omega + \int_{\Gamma_j} \left(\frac{\partial \bar{u}_h}{\partial n} - \frac{\partial u_h}{\partial n} \right) v d\Gamma$$

$$u_{0tol} = 5.1192x$$

$$u_{1to2} = 5.1192(12 - x) + 8.2944(-6 + x)$$

$$\Delta u_{0tol} = 0$$

$$\Delta u_{1to2} = 0$$

Finding the residuals; $R_j = f + \Delta u_h$

$$R_1 = \frac{x^3}{1000} = R_2$$

$$\frac{1}{6}\begin{bmatrix} 1 & -1 \\ -1 & 1 \end{bmatrix}\begin{bmatrix} \bar{e}_{11} \\ \bar{e}_{12} \end{bmatrix} = \begin{bmatrix} \int_0^6 R_1 \cdot \left(\frac{6-x}{6}\right)dx \\ \int_6^{12} R_1 \cdot \left(\frac{x}{6}\right)dx \end{bmatrix} + \begin{bmatrix} \left(\dfrac{\dfrac{\partial u_{0to1}}{\partial x}+\dfrac{\partial u_{1to2}}{\partial x}}{2}\right)-\dfrac{\partial u_{0to1}}{\partial x} \\ \left(\dfrac{\dfrac{\partial u_{0to1}}{\partial x}+\dfrac{\partial u_{1to2}}{\partial x}}{2}\right)-\dfrac{\partial u_{0to1}}{\partial x} \end{bmatrix}$$

$$\frac{1}{6}\cdot\begin{bmatrix} 1 & -1 \\ -1 & 1 \end{bmatrix}\begin{bmatrix} \bar{e}_{11} \\ \bar{e}_{12} \end{bmatrix} = \begin{bmatrix} 0.0648 \\ 0.2592 \end{bmatrix} + \begin{bmatrix} -0.972 \\ -0.972 \end{bmatrix}$$

as per the boundary condition, $u=0$ at $x=0$, therefore, $\bar{e}_{11}=0$.

$$\bar{e}_{12}=-4.2768$$

now,

$$\frac{1}{6}\cdot\begin{bmatrix} 1 & -1 \\ -1 & 1 \end{bmatrix}\begin{bmatrix} \bar{e}_{21} \\ \bar{e}_{22} \end{bmatrix} = \begin{bmatrix} 1.6848 \\ 3.1752 \end{bmatrix} + \begin{bmatrix} 0.972 \\ 0.972 \end{bmatrix}$$

$$\bar{e}_{22}=24.8832$$

$$\|e_j\|=29.16$$

7.7.4 Error by Goal Oriented Estimator

Let the quantity of interest (QoI) be the average displacement between node 2 and 3:

$$f_0 = Q(u_h) = \left\{\left(\frac{1}{6}\right)\times\int_6^{12} u_{2to3}dx\right\} = 40.2408$$

Assuming it as a force vector and solving the dual problem, we get

$$\begin{bmatrix} z_1 \\ z_2 \\ z_3 \end{bmatrix} = \begin{bmatrix} 0 \\ 2173 \\ 2897.34 \end{bmatrix}$$

The interpolation function of dual problem can be expressed as

$$z_{1to2}=362.167x$$

$$z_{2to3}=362.167(12-x)+482.89(-6+x)$$

Due to linear interpolation function

$$\Delta z_{1to2} = \Delta z_{2to3} = 0$$

Residuals for dual problem $[(R_j)_z = f_0 + \Delta z_h]$:

$$(R_1)_z = f_0 = (R_2)_z$$

$$\|(R_1)_z\|_{L^2(\Omega_1)} = \left[\int_0^6 \{(R_1)_z\}^2 \, dx \right]^{\frac{1}{2}} = 98.5694$$

$$\|(R_2)_z\|_{L^2(\Omega_2)} = \left[\int_6^{12} \{(R_2)_z\}^2 \, dx \right]^{\frac{1}{2}} = 98.5694$$

Jump at the boundary:

at inter-element boundary, node 2, $(J_1)_z = \nabla z_{1to2} - \nabla z_{2to3} = 241.445$

at Neumann boundary, node 3, $(J_2)_z = 0 - \nabla z_{2to3} = -120.772$

Therefore,

$$\omega_1 = \left[h_1^2 \|(R_1)_z\|_{L^2(\Omega_1)}^2 + h_1 \left(\frac{(J_1)_z}{2} \right)^2 \right]^{\frac{1}{2}} = 661.224$$

$$\omega_2 = \left[h_2^2 \|(R_2)_z\|_{L^2(\Omega_2)}^2 + h_2 \left\{ \left(\frac{(J_1)_z}{2} \right)^2 + ((J_2)_z)^2 \right\} \right]^{\frac{1}{2}} = 724.334$$

Therefore, error on first element $= \eta_1.\omega_1 = 1762.9$, error on first element $= \eta_2.\omega_2 = 11429.7$ and total error

$$|Q(e_h)| = \eta_1.\omega_1 + \eta_2.\omega_2 = 13192.7.$$

This also shows large error in element 2.

7.7.5 Error by Hierarchical and Wavelet Based Estimator

In this case, we use the difference of FEM solution at two consecutive levels. Finding two FEM solutions to compute the total error indicator will be unadvisable. Therefore, we have to use some approximate method which uses only one set of FEM solution to find the two levels of approximate difference. This approximate difference may be a good error indicator.

One method to implement the approach is to use FEM solution and its coarse solution which we obtained using lazy wavelets. In this problem, we used B-spline lazy wavelets[8]. For eight elements of equal sizes, the finite element solution is obtained as

$$\{U_8\} = \begin{bmatrix} 0 & 7.77 & 15.54 & 23.24 & 30.71 & 37.69 & 43.7 & 48.05 & 49.76 \end{bmatrix}$$

This solution is transformed two times using lazy wavelets. The estimated two scale error (using the norm as $U = a(u,u)$) is shown below:

$$\|U_8 - U_4\| = \begin{bmatrix} 2.2 \times 10^{-5} & 2.2 \times 10^{-5} & 7.8 \times 10^{-3} & 7.8 \times 10^{-3} & 0.16 & 0.16 & 1.15 & 1.15 \end{bmatrix}$$

$$\|U_4 - U_2\| = \begin{bmatrix} 1.1 \times 10^{-2} & 1.1 \times 10^{-2} & 3.997 & 3.997 \end{bmatrix}$$

The largest two scale errors $\|U_4 - U_2\|$ are in two elements $(x = 6\,\text{to}\,9)$ and $(x = 9\,\text{to}\,12)$ but the largest two scale errors in $\|U_8 - U_4\|$ are in two elements $(x = 9\,\text{to}\,10.5)$ and $(x = 10.5\,\text{to}\,12)$. This shows the trend of reduction in two scale error with refinement and suggests the elements to be refined. We found this approach suitable to many problems. It can be implemented on the non-uniform grid also.

REFERENCES

1. T. Grätsch and K.J. Bathe (2005), A posteriori error estimation techniques in practical finite element analysis. *Computers &Structures*, Vol. 83, pp. 235–265.
2. M. Ainsworth and J.T. Oden (2000), *A Posteriori Error Estimation in Finite Element Analysis*, Wiley, New York.
3. J. E. Akin (2005), *Finite Element Analysis with Error Estimators: An Introduction to the FEM and Adaptive Error Analysis for Engineering Students*, Butterworth-Heinemann, Amsterdam, the Netherlands.
4. O. C. Zienkiewicz, R. L. Taylor and J.Z. Zhu (2005), *The Finite Element Method: Its Basis & Fundamentals*, Butterworth-Heinemann, Amsterdam, the Netherlands.
5. J. T. Oden, and S. Prudhomme (2001), Goal-oriented error estimation and adaptivity for the finite element method. *Computers &Mathematics with Applications*, Vol. 41, No. (5–6), pp. 735–756.
6. R. E. Bank and R. K. Smith (1993), A posteriori error estimates based on hierarchical bases. *SIAM Journal on Numerical Analysis*, Vol. 30, pp. 921–935.
7. R. Sudarshan, K. Amaratunga and T. Grätsch (2006), A combined approach for goal-oriented error estimation and adaptivity using operator-customized finite element wavelets. *International Journal for Numerical Methods in Engineering*, Vol. 66, pp. 1002–1035.
8. E. J. Stollnitz, T. D. DeRose and D. H. Salesin (1996), *Wavelets for Computer Graphics*, Morgan Kaufmann, San Francisco, CA.

Appendix A: Sets

A SET OR "ANY COLLECTION of objects" is the most fundamental notion in mathematics. Engineers and often scientists do not use set theory frequently in their practice; therefore, they may find this chapter useful. Rigorous thinking of mathematics can be easily and accurately expressed with the help of sets and relations. To help quick reference of notations of set theory, we are presenting a list of common notations.

A.1 SOME NOTATIONS

$\in$ "a member of" e.g., $a \in A$ means a is member of A

$\notin$ "not a member of" e.g., $a \notin A$ means a is not a member of A

$\subseteq$ "subset" e.g., $B \subseteq A$ means B is subset of A or B is contained in A

$\supseteq$ "subset" e.g., $B \supseteq A$ means B contains A

$\subset$ "proper subset" e.g., $B \subset A$ means $B \subseteq A$ and $B \neq A$

$\supset$ "proper subset" e.g., $B \supset A$ means $B \supseteq A$ and $B \neq A$

$\cup$ "union" e.g., $A \cup B$ means set with element that belongs to A or B

$\cap$ "intersection" e.g., $A \cap B$ means set with element that belongs to A and B

$-$ or $\backslash$ "difference" e.g., $A - B$ or $A \backslash B$ means set of elements that belongs to A but not to B

Δ "symmetric difference" e.g., $A \Delta B$ means set of elements that belongs to A and B but not to $A \cap B$

$\otimes$ "tensor product" e.g., $\otimes$ Let $u = \{u_1, u_2, \ldots\}$ and $v = \{v_1, v_2, \ldots\}$ then

$$u \otimes v = \begin{bmatrix} u_1 v_1 & u_1 v_2 & \cdot \\ u_2 v_2 & u_2 v_2 & \cdot \\ \cdot & \cdot & \cdot \end{bmatrix}$$

$\Rightarrow$ "implies" e.g., $A \Rightarrow B$ means A implies B or if A then B

$\Leftrightarrow$ "iff" e.g., $A \Leftrightarrow B$ means A if and only if B or A iff B

$\forall$ "for all" e.g., $\forall x$ means for all x

$\exists$ "there exists" e.g., $\exists x$ means there exists x

$\varnothing$	denotes empty set or set containing no element
$\mathbb{C}$	denotes the set of all complex numbers
$\mathbb{N}$	denotes the set of all natural numbers or positive integers*
$\mathbb{Q}$	denotes the set of all rational numbers i.e., n/m where n and m are integers and $m \neq 0$
$\mathbb{R}$	denotes the set of all real numbers
$\mathbb{Z}$	denotes the set of all integers numbers or $\{\ldots -2, -1, 0, 1, 2 \ldots\}$

* **Note:** There is no universal agreement regarding inclusion of zero in the set of natural numbers.

A.2 SETS

We define a set as an entity made of the gathering of mathematical elements a, b,... called constants. The set is specified by listing its elements within curly brackets. For example $\{1,3,5,7\}$ denotes a set with elements 1,3, etc. We shall denote sets by the uppercase letter A, B... and their elements by lowercase letters a, b,... The statement "the element a belongs to the set A" will be written as $a \in A$.

A much more general approach is to define set A as the collections of all elements x that satisfy a property P; we indicate this by writing $A = \{x \mid x \text{ satisfy } P\}$. We read it as "$A$ is the set of all x such that x satisfies P". Some books use colon ":" instead of vertical bar "|" as a symbol for "such that".

A set whose elements are all sets is called a collection or family of set. Script letters such as $\mathcal{A}, \mathcal{M}, \mathcal{U}$ are often used for these collections. For example, $\mathcal{A} = \{A_1, A_2, \ldots A_n\}$.

Using the interval notation, we have the real line $\mathbb{R} = (-\infty, \infty)$, closed interval $[a,b] = \{x \mid a \leq x \leq b\}$, open interval $(a, b) = \{x \mid a < x < b\}$ and half open interval and half closed interval $[a, b) = \{x \mid a \leq x < b\}$ and $(a, b] = \{x \mid a < x \leq b\}$.

For family of infinitely many sets A_λ, $\lambda \in \Lambda$ (Λ is called the index set), union and intersection are defined as

$$\bigcup A_\lambda = \{x \mid x \in A_\lambda \text{ for some } \lambda \in \Lambda\} \text{ and}$$

$$\bigcap A_\lambda = \{x \mid x \in A_\lambda \text{ for all } \lambda \in \Lambda\}$$

We defined the relative complement of B in A, also called difference set and denoted by $A - B$ defined as

$$A - B = \{x \mid x \in A \text{ and } x \notin B\}$$

The symmetric difference of A and B denoted by $A\Delta B$ is defined as

$$A\Delta B = (A-B)\cup(B-A)$$

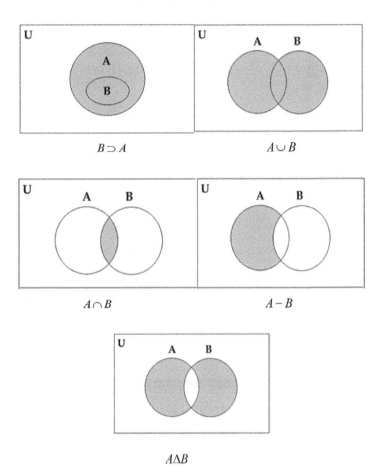

$B\supset A$

$A\cup B$

$A\cap B$

$A-B$

$A\Delta B$

Definition: A subset S of F is called bounded above if there exist an element $b\in F$ such that $x\le b$ whenever $x\in S$ and any such b is called an upper bound of S.

Example A.1:

Suppose a set $S = \{\sum_n \frac{1}{2^n} \mid n = 0,1,2,3,\ldots\ldots\}$ then 2 is an upper bound of S. Similarly we can define bounded below set for which there exist an element $a\in F$ such that $x\ge a$ whenever $x\in S$ and any such a is called a lower bound of S.

Definition: A set is said to be bounded if it is both bounded above and bounded below.

Supremum (or least upper bound): A number m is called supremum of a set A if

(i) m is an upper bound of A; (ii) for any small positive number ε, m-ε is not an upper bound. Generally, supremum of a set A is denoted as sup(A).

Similarly we can define **Infimum** of a set. Infimum (or greatest lower bound): A number n is called infimum of a set A if (i) n is a lower bound of A and (ii) for any small positive number ε, m$-\varepsilon$ is not a lower bound. Generally, infimum of a set A is denoted as inf(A).

Note: Supremum or infimum of a set may or may not belong to the set. But, if sup(A) exists and belongs to A, it is the largest member of A and is also denoted as max(A), the maximum of A. If inf(A) exists and belongs to A then it is denoted as min(A).

Example A.2:

$A = \{0 \le r \in Q \backslash r^2 < 2\}$

Here is a least upper bound of A, i.e. sup(A) but max(A).

Definition (Product of sets and equivalence relation): Let us consider a set A and Γ a part of $A \times A$. Given an element (a; b) of Γ, we use the notation aRb. We say that Γ defines an equivalence relation if the following three conditions are satisfied:

R is reflexive: for any $a \in A$, aRa.
    *i.*e., if R is any relation then a is related to a itself.
R is symmetric: for any $(a; b) \in A \times A$, aRb implies bRa.
    i.e., if a is related to b then b is also related to a.
R is transitive: for any $(a; b; c) \in A \times A \times A$, (aRb and bRc) implies aRc.
    i.e., if a is related to b and b is related to c, then a is related to c also.

Example A.3:

1. Let A is a set of all straight lines, $a, b \in A$,
   a. Let R_1 be a relation between elements of A such that for all $a, b \in A$, aR_1b means a is perpendicular to b.
   Then R_1 is symmetric but neither transitive nor reflexive.
   b. Let R_2 be a relation between elements of A such that for all $a, b \in A$, aR_2b means a is parallel to b. Then this R_2 is reflexive, symmetric and transitive, and $(A \times A, R_2)$ forms an equivalence relation.
2. Let $A = \{1,2,3\}$

$$A \times A = \{(1,1),(1,2),(1,3),(2,1),(2,2),(2,3),(3,1),(3,2),(3,3)\}$$

a. If the relation aRb is defined as $a = b$ then (Γ, R) forms an equivalence relation for all $(a, b) \in \Gamma \subset A \times A$ where

$$\Gamma = \{(1,1),(2,2),(3,3)\}.$$

b. If aRb is defined as $a < b$ then R is transitive but not reflexive and symmetric; therefore, R will not form an equivalence relation.
In such relation $\Gamma = \{(1,2),(2,3),(1,3)\}$.

Appendix B: Fields

A FIELD IS AN ALGEBRAIC structure with notions of addition and multiplication satisfying certain axioms. The most commonly used fields are the field of real numbers, the field of complex numbers and the field of rational numbers, but there are also finite fields, fields of functions, various algebraic number fields and so forth. Any field may be used as the scalars for a vector space, which is the standard general context for linear algebra.

Definition: A field is a set K, containing at least two elements, on which two operations $+$ and $\cdot$ (called addition and multiplication, respectively) are defined so that for each pair of elements x, y in K

1. Addition is commutative,

   $x + y = y + x$ for all x and y in K.

2. Addition is associative,

   $x + (y + z) = (x + y) + z$ for all x, y and z in K.

3. There is a unique element 0 (Zero) in K such that $x + 0 = x$, for every x in K.

4. To each x in K there corresponds a unique element $(-x)$ in K such that $x + (-x) = 0$.

5. Multiplication is commutative,

   $x \cdot y = y \cdot x$ for all x and y in K.

6. Multiplication is associative,

   $x(yz) = (xy)z$ for all x, y and z in K.

7. There is a unique non-zero element 1 (one) in K such that

   $x1 = x$, for every x in K.

8. To each non-zero x in K there corresponds a unique element

   x^{-1} in K such that $x \cdot x^{-1} = 1$.

9. Multiplication distributes over addition; that is,

   $x(y + z) = xy + xz$, for all x, y and z in K.

Example B.1:

Rational numbers can be written as fractions $\frac{a}{b}$, where a and b are integers, and $b \neq 0$. So the set of rational number forms a field with the usual operation of addition and multiplication. The additive inverse of such a fraction is $-\frac{a}{b}$, and the multiplicative inverse (provided that $a \neq 0$) is $\frac{b}{a}$.

Example B.2:

The real numbers R, with the usual operations of addition and multiplication, also form a field.

Example B.3:

The set of complex numbers, denoted C, is the set of ordered pairs of real numbers (a, b), with the operations of addition and multiplication defined by: (a, b) + (c, d) = (a + c, b + d) and (a, b) · (c, d) = (ac − bd, ad + bc). Note that (0, 1) · (0, 1) = (−1, 0), so (0, 1) is a complex number whose square is −1. We usually write i for (0, 1), and a + ib or a + bi for (a, b). In that case, the multiplication is (a + ib)(c + id) = ac − bd + i(ad + bc). The real numbers a and b are called the real and imaginary parts of a + ib, respectively. So, C forms a field with the above multiplication and addition.

> **Finite Fields.** Finite fields (also called Galois fields) are fields with finitely many elements.
>
> **Remark.** In a finite field there is necessarily an integer n such that $1 + 1 + \ldots + 1$ (n repeated terms) equals 0. It can be shown that the smallest such n must be a prime number, called the characteristic of the field. If a (necessarily infinite) field has the property that $1 + 1 + \ldots + 1$ is never zero, for any number of summands, such as in Q. In such cases, the characteristic is said to be zero.

Example B.4:

A basic class of finite fields are the fields Z_p with p elements (p a prime number): $Z_p = \frac{Z}{pZ} = \{0,1,2,3,\ldots p-1\}$, where the operations are defined by performing the operation in the set of integers Z, dividing by p and taking the remainder.

In particular if we take $p = 2$ then, $Z_2 = \{0,1\}$ and with the addition and multiplication modulo 2, it forms a field.

$$0+0=0, \quad 0+1=1, \quad 1+0=1, \quad 1+1=0$$

$$0.0=0, \quad 0.1=0, \quad 1.0=0, \quad 1.1=1.$$

Theorem: Let K be a field. (1) The additive identity in K is unique. (2) The additive inverse of an element of K is unique. (3) The multiplicative identity of K is unique. (4) The multiplicative inverse of a non-zero element of K is unique.

Theorem: (Cancellation Laws). Let a, b and c be elements of a field F.
(1) If $a + b = b + c$, then a = c. (2) If $ab = bc$ and $b \neq 0$, then a = c.

Theorem: Let a and b be elements of a field K. Then (1) $a \cdot 0 = 0$ (2) $(-a)b = a(-b) = -ab$ (3) $(-a)(-b) = ab$.

Application of fields in discrete mathematics: In this section, we will solve the linear system of equations for the unknown variables in Z_2.

Consider the linear system of equation.

$$x_1 + x_3 + x_5 = 1$$

$$x_1 + x_2 + x_4 + x_5 = 0$$

$$x_1 + x_4 = 1$$

We can solve this system using Gaussian elimination, using the fact all operations in modular arithmetic. In particular, $x, y \in Z_2$, then $x + x = 0$, and $x + y = 0$ implies $x = y$. Adding the first equation to the second and third equation, we get

$$x_1 + x_3 + x_5 = 1$$
$$x_2 + x_3 + x_4 = 1$$
$$x_3 + x_4 + x_5 = 0$$

Now, adding the third equation to the first and second equation, we get

$$x_1 + x_4 = 1$$
$$x_2 + x_5 = 1$$
$$x_3 + x_4 + x_5 = 0$$

So we have

$$x_1 = 1 + x_4$$
$$x_2 = 1 + x_5$$
$$x_3 = x_4 + x_5$$

The variable x_4 and x_5 can be take any values; suppose $x_4 = a$ and $x_5 = b$.
We get the solution

$$x = (1 + a, 1 + b, a + b, a, b) = (1,1,0,0,0) + a(1,0,1,1,0) + b(0,1,1,0,1).$$

Here (1,0,1,1,0) and (0,1,1,0,1) are called as complementary solution, and (1,1,0,0,0) is called as particular solution.

Appendix C: C^m and L^p Spaces

O UR UNDERSTANDING OF AN article depends on its continuity because it helps in imag- ination and establishing proper correlations among the pre-existing concepts. It is not only the requirement of mind for imagination, etc. Some sort of continuity is essential for progress in every field. Discontinuity not only irritates but also creates serious doubts on the validity of the concept. It will be hard to accept discontinuity in the field variable of differential equations. In mathematics, continuity is very well formulized and the concept of weak continuity is very helpful in searching the solution in the wider space.

A few decades ago, the discontinuities introduced by engineers to solve the structural problems using the finite element method surprised mathematicians. Breaking of func- tion in the potential energy expression (equations 1.14, 1.15, etc.) where one expression of field variable is replaced by many polynomials generated serious doubts in the minds of mathematicians about its validity. First, engineers realized the need to accommodate some discontinuities for their complex domain boundary value problems, and then math- ematicians proved how some discontinuities are not problems and in fact are a different kind of continuity. The knowledge of another space which accepts more than the classical continuity is very useful for understanding the precise mathematical background of the finite element method and underlines some limitations of the method. To understand the reason behind the success and some possible failures, we have to understand the notation of continuous spaces such as C^m, L^p and Sobolev spaces H^m.

Since we are interested to learn about the continuity of spaces, it is necessary to know what do we mean by continuity. Continuity is defined in terms of distance function or metric as follows:

Definition (Metric space): Let X be a non-empty set. A metric on X is a function $d: X \times X \to \mathbb{R}$ satisfying the following axioms $\forall x, y, z \in X$:

1. $d(x, y) \geq 0$ and $d(x, y) = 0 \Leftrightarrow x = y$ identity

2. $d(x, y) = d(y, x)$ symmetry

3. $d(x, y) \leq d(x, z) + d(y, z)$ triangular inequality

Then d is called a **"metric"** on X. The set X together with the metric d, denoted by (X, d) is called **metric space**. The symbol $\times$ denotes the Cartesian product of two sets. The real num- ber $d(x, y)$ is the distance between x and y in X. It can be noted that d is a non-negative, single, real valued function.

Example C.1:

1. Define $d : X \times X \to \mathbb{R}$ such that

$$d(x, y) = 0, \quad x = y$$

$$1, \quad x \neq y$$

Then (X, d) is a metric space.

2. The set of real numbers $\mathbb{R}$ with distance function defined as

$$d(x, y) = |x - y|, \text{ for } x, y \in \mathbb{R}$$

forms a metric space.

3. The set $\mathbb{R}^n$ of ordered n-tuples of real numbers $x = (x_1, x_2, \ldots \ldots x_n)$
   a. with distance function

$$d(x, y) = \left\{ \sum_{k=1}^{n} (y_k - x_k)^2 \right\}^{1/2}$$

is a metric space which we call Euclidean n-space $\mathbb{R}^n$.
   b. with the distance function

$$d(x, y) = \max\{|y_i - x_i| : 1 \leq i \leq n\}$$

also forms a metric space.

4. The set $C[a, b]$ of all continuous real-valued functions defined on interval $[a,b]$
   a. with distance function

$$d(f, g) = \sup_{a \leq x \leq b} \{|g(x) - f(x)|\}$$

forms a metric space.
   b. with the distance function

$$d(f, g) = \left[\int_a^b \{g(x) - f(x)\}^2 \, dx \right]^{1/2}$$

also forms a metric space.

Definition (Continuous functions in metric space): Let (X, d_X) and (Y, d_Y) be metric spaces, and let $f : X \to Y$ be a function. The function f is called continuous at $p \in X$ if for every $\epsilon > 0$ there is a $\delta > 0$ such that

$$d_X(x, p) < \delta \Rightarrow d_Y(f(x), f(p)) < \epsilon$$

The function f is continuous at x if it is continuous at every point in X.

Example C.2:

Let us check the continuity of the following function

$$f(x) = \begin{cases} 1 & x < 2 \\ 1/2 & x = 2 \\ -1 & x > 2 \end{cases}$$

Let $\epsilon = 1/10$ and $p = 2$

$$d_Y\big(f(x), f(p)\big) < \epsilon$$

or

$$\frac{1}{2} - \epsilon < f(x) < \frac{1}{2} + \epsilon$$

or

$$\frac{1}{2} - \frac{1}{10} < f(x) < \frac{1}{2} + \frac{1}{10}$$

or

$$\frac{4}{10} < f(x) < \frac{6}{10}$$

But we do not have any δ where this condition is satisfied; therefore, the function is discontinuous.

C.1 C^m SPACES

The function assigns some real (or complex) number for each $x \in \Omega \subset \mathbb{R}^d$, i.e., $f : \Omega \to \mathbb{R}$ or $\mathbb{C}$. If we collect all continuous functions $f(x)$ in some domain and form a set, we call it a functional space. Here Ω is an open set whose boundary is denoted by $\Gamma = \partial\Omega$. The closure of a set (open set with boundary) is represented as $\bar{\Omega}$.

C.1.1 Multi-dimensional Spaces and Multi-index Notation

In this section and subsequent sections we will deal with partial derivatives of all orders; therefore, it is very useful to introduce multi-index notation of higher derivatives. Consider a multi-dimensional domain $x \in \mathbb{R}^n$. The multi-index notation for the partial differential equation of function $u(x)$ may be expressed as

$$D^\alpha u(x) = \frac{\partial^{|\alpha|} u}{\partial x_1^{\alpha_1} \partial x_2^{\alpha_2} \ldots \partial x_n^{\alpha_n}}$$

$\alpha = (\alpha_1, \alpha_2 \ldots, \alpha_n) \in \mathbb{N}^n$ (Non-negative integer) and the sum is denoted by $|\alpha|$

$$|\alpha| = \alpha_1 + \alpha_2 \ldots + \alpha_n$$

Example C.3:

Consider a partial differential operator L. It can be expressed using compact notation as

$$L = \sum_{|\alpha| \le m} a_\alpha D^\alpha u$$

where a_α are coefficients. For dimension of space $n = 3$ and order of derivative $m = 2$, it can be expressed as

$$L = \sum_{|\alpha|=0} a_\alpha D^\alpha u + \sum_{|\alpha|=1} a_\alpha D^\alpha u + \sum_{|\alpha|=2} a_\alpha D^\alpha u.$$

$$L = \left(a_{000} \frac{\partial^0 u}{\partial x_1^0 \partial x_2^0 \partial x_3^0} \right) + \left(a_{100} \frac{\partial^1 u}{\partial x_1^1 \partial x_2^0 \partial x_3^0} + a_{010} \frac{\partial^1 u}{\partial x_1^0 \partial x_2^1 \partial x_3^0} + a_{001} \frac{\partial^1 u}{\partial x_1^0 \partial x_2^0 \partial x_3^1} \right) +$$

$$\left(a_{110} \frac{\partial^2 u}{\partial x_1^1 \partial x_2^1 \partial x_3^0} + a_{011} \frac{\partial^2 u}{\partial x_1^0 \partial x_2^1 \partial x_3^1} + a_{101} \frac{\partial^2 u}{\partial x_1^1 \partial x_2^0 \partial x_3^1} + \right.$$

$$\left. a_{200} \frac{\partial^2 u}{\partial x_1^2 \partial x_2^0 \partial x_3^0} + a_{020} \frac{\partial^2 u}{\partial x_1^0 \partial x_2^2 \partial x_3^0} + a_{002} \frac{\partial^2 u}{\partial x_1^0 \partial x_2^0 \partial x_3^2} \right)$$

$$L = a_{000} u + a_{100} \frac{\partial u}{\partial x_1} + a_{010} \frac{\partial u}{\partial x_2} + a_{001} \frac{\partial u}{\partial x_3} + a_{110} \frac{\partial^2 u}{\partial x_1 \partial x_2}$$

$$+ a_{011} \frac{\partial^2 u}{\partial x_2 \partial x_3} + a_{101} \frac{\partial^2 u}{\partial x_1 \partial x_3} + a_{200} \frac{\partial^2 u}{\partial x_1^2} + a_{020} \frac{\partial^2 u}{\partial x_2^2} + a_{002} \frac{\partial^2 u}{\partial x_3^2}$$

With such simplified expressions of multi-index notation, we can readily represent partial derivatives of any order m for any domain $\Omega \subset \mathbb{R}^n$.

C.1.2 Space of Continuously Differentiable Functions

Consider the examples of functional spaces $C^m(\Omega)$. The C^m space can be defined as

$$C^m(\Omega) = \left\{ u(x_1, x_2, \ldots x_n) \middle| D^\alpha u \in C^0(\Omega), \forall |\alpha| = m \right\}$$

Example C.4:

If $\Omega \subset \mathbb{R}^1$, for the non-negative integer m, the class C^m consists of all continuous functions u whose derivatives $\frac{du}{dx}, \frac{d^2 u}{dx^2}, \frac{d^3 u}{dx^3}, \ldots \frac{d^m u}{dx^m}$ exist but the derivative of $\frac{d^m u}{dx^m}$ does not exist; in other words, $\frac{d^m u}{dx^m}$ has C^0 continuity.

Similarly, if $u \in C^3(\Omega)$ and $\Omega \subset \mathbb{R}^2$ then $u, u_x, u_y, u_{xx}, u_{xy}, u_{yy}, u_{xxx}, u_{xxy}, u_{xyy}$ and u_{yyy} are continuous on Ω. Figure C.1 shows examples of functions in different continuous spaces.

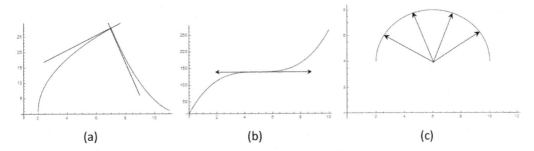

$$(a) \qquad\qquad (b) \qquad\qquad (c)$$

FIGURE C.1   Continuity of the functions for domain $\Omega \subset \mathbb{R}^1$ (a) C^0, (b) C^1, (c) C^2.

The distance between two functions $u, v \in C^0(\Omega)$ is defined as

$$d(u,v) = \max_{x \in \Omega} |u(x) - v(x)|$$

It satisfies the following axioms:

$$d(u,v) \geq 0 \text{ and, } d(u,v) = 0 \text{ iff } u \equiv v;$$

$d(u, v + w) \leq d(u,v) + d(u,w)$, the triangle inequality.
   This space is called a metric space with metric (distance) d
   A norm in $C^0(\Omega)$ can be defined as

$$\|u(x)\| = \max_{x \in \Omega} |u(x)|$$

Here $\|u(x)\| = d(u,v)$ for $v = 0$
   In $C^m(\Omega)$, the norm is defined as

$$\|u(x)\| = \max_{0 \leq \alpha \leq m} \max_{x \in \Omega} |D^\alpha u|$$

The $C^m(\Omega)$ space is linear because for any real numbers α and β and $u, v \in C^m(\Omega)$ we get $\alpha u + \beta v \in C^m(\Omega)$. It can be verified easily that

$$C^\infty \subset \ldots \subset C^m \subset \ldots C^1 \subset C^0$$

$C^\infty(\Omega)$ is the space of infinitely differentiable functions. For example, $\sin x, e^x \in C^\infty$.

C.2 MEASURE THEORY AND L^p SPACES

Measure theory is essential to understand Lebesgue space, which is a Hilbert space. Therefore we are presenting a brief and essential part of measure theory. Measure theory finds wide application in the area of probability, functional analysis and dynamical systems etc. Discussion on various measure theories such as probability measure, Jordan measure, etc. is beyond the scope of this book, but we will focus on general theory to understand Lebesgue measure.

Definition: A collection $\mathcal{M}$ of subsets of a set $\mathcal{X}$ is called σ algebra of the set $\mathcal{X}$ if

1. null set $\varnothing \in \mathcal{M}$
2. a set $A \in \mathcal{M}$ then its complement $A^c = \mathcal{X}\!/_A \in \mathcal{M}$
3. $\{A_i | i \in \mathbb{N}\}$ is a countable family of sets in $\mathcal{M}$ then $U_{i=1}^{\infty} A_i \in \mathcal{M}$

The pair $(\mathcal{X}, \mathcal{M})$ is known as a measurable space because it tells which subsets are measurable. It follows from the definition that whole space $\mathcal{X}$ is measurable and the intersection of any pair of measurable sets is measurable.

Definition: For a measurable space $(\mathcal{X}, \mathcal{M})$, a measure μ on a set $\mathcal{X}$ is function $\mu : \mathcal{M} \rightarrow [0, \infty]$ such that

1. $\mu(A) \geq 0$ for all measurable sets A and $\mu(\varnothing) = 0$
2. if $A_i \in \mathcal{M}$ for $i \in \mathbb{N}$ is any finite or countably infinite mutually disjoint sets in $\mathcal{M}$, then $\mu(U_{i=1}^{\infty}) = \sum_{i=1}^{\infty} \mu(A_i)$

A measure space is triple $(\mathcal{X}, \mathcal{M}, \mu)$ consisting of a set $\mathcal{X}$, a σ algebra $\mathcal{M}$ on $\mathcal{X}$ (in other words measurable space $(\mathcal{X}, \mathcal{M})$ together with a measurable μ). Measure can be signed measures, complex measures and vector-valued measures, but these measures will not be required here.

Lebesgue measure $\mu(A)$ is a very natural generalization of the concept of length, area and volume. The other examples of Lebesgue measures include the increment $f(b) - f(a)$ of a Shift $f(t)$ downward an interval [a, b], the integral of a function $f(t)$ downward an interval [a, b] or the integral of a non-negative function over $\Omega \subset \mathbb{R}^d (d = 1,2,3... \text{ etc.})$.

If $\mathcal{M}$ is the collection of rectangles which are defined by closed rectangles $(a \leq x \leq b)$ and $(c \leq y \leq d)$ or open rectangles $(a < x < b)$ and $(c < y < d)$ then the measure of a close, open or half open rectangle is equal to $(b-a)(d-c)$. This quantity is generalized on $\mathbb{R}^n$ and defines Lebesgue measure μ on $\mathbb{R}^n$ as a quantity or function on σ algebra such that

$$\mu(A) = (b_1 - a_1)(b_2 - a_2)........(b_n - a_n)$$

Theorem: A subset A of $\mathbb{R}^n$ is a Lebesgue measure if and only if there exists $\epsilon > 0$ for outer measure of an open set O and inner measure of closed and bounded set C of a set A such that $\mu(\%_c)$ or $\mu(o - c) < \epsilon$. In other words, set A is a Lebesgue measure if $\mu(o) = \mu(c)$.

Thus the approximate value of outer and inner measures is called Lebesgue measure $\mu(A)$.

Definition (Sets of measure zero): Let $(\mathcal{X}, \mathcal{M}, \mu)$ be a measure space. A subset A of $\mathcal{X}$ is said to have measure zero if $\mu(A) = 0$. For example, a measure of a curve is its length, but the length of a point is zero.

Suppose f is a function defined in $\mathbb{R}^d (d = 1, 2, 3)$. We are approximating f with the help of $\sum g_i$ whose measures are equal almost everywhere except at a number of isolated points in $\mathbb{R}^1$ or curve in $\mathbb{R}^2$ or surfaces in $\mathbb{R}^3$, then $\mu(f)$ is equal to $\mu(\sum g_i)$ a.e. A property that holds except on a set of measure zero is said to hold almost everywhere (abbreviated a.e.).

C.3 L^p SPACES

Theorem: $L^p(R)$ is a vector space.

Proof. For any $f, g \in L^p(R)$ and $\alpha, \beta \in F$, we define for any $x \in R$

$$(f + g)(x) = f(x) + g(x)$$

$$(\alpha f)(x) = \alpha f(x)$$

1.
   a. The addition is commutative, i.e.,

   $$g + f = f + g$$

   b. The addition is associative, i.e.,

   $$(f + g) + h = f + (g + h)$$

   c. There exists zero map $z = 0 \in L^p(R)$ such that

   $$g + 0 = g$$

   0 is the identity element for addition.
   d. For each $f \in L^p(R)$ there exists $-f \in L^p(R)$ such that

   $$f - f = 0$$

2.
   a. Distributive of scalar multiplication with respect to vector addition

   $$\alpha \cdot (f + g) = \alpha \cdot f + \alpha \cdot g$$

   b. Distributive of scalar multiplication with respect to field addition

   $$(\alpha + \beta) \cdot f = \alpha \cdot f + \beta \cdot f$$

c. Identity element of scalar multiplication

$$1 \cdot f = f$$

d. Scalar multiplication with field multiplication

$$(\alpha \cdot \beta) \cdot f = \alpha \cdot (\beta \cdot f)$$

Hence all the axioms of vector space are satisfied.

Theorem: The space $L^p(R)$ is Banach space.

Proof. Let $\{g_n\}_{n=1}^{\infty}$ be a Cauchy sequence in $L^p(R)$, consider a subsequence $\{g_{n_i}\}_{i=1}^{\infty}$ of $\{g_n\}_{n=1}^{\infty}$ such that

$$\left\| g_{n_{i+1}} - g_{n_i} \right\| \le 2^{-i} \text{ for all } i \ge 1.$$

Now consider the series

$$g(x) = g_{n_1}(x) + \sum_{i=1}^{\infty} \left(g_{n_{i+1}}(x) - g_{n_i}(x) \right)$$

and

$$h(x) = \left| g_{n_1}(x) \right| + \sum_{i=1}^{\infty} \left| \left(g_{n_{i+1}}(x) - g_{n_i}(x) \right) \right|,$$

Together the partial sums

$$s_I(g)(x) = g_{n_1}(x) + \sum_{i=1}^{I} \left(g_{n_{i+1}}(x) - g_{n_i}(x) \right)$$

and

$$s_I(h)(x) = \left| g_{n_1}(x) \right| + \sum_{i=1}^{I} \left| \left(g_{n_{i+1}}(x) - g_{n_i}(x) \right) \right|,$$

Now by using triangle inequality

$$\left\| s_I(h) \right\| \le \left\| g_{n_1} \right\| + \sum_{i=1}^{I} \left\| g_{n_{i+1}} - g_{n_i} \right\|$$

$$\leq \|g_{n_1}\| + \sum_{i=1}^{I} 2^{-i}$$

Taking limit $I \to \infty$ and applying the monotone convergence theorem proves that $\int |h|^2 dx \leq \infty$, and since $|g| \leq h$, we have $g \in L^p(R)$. So the series defining g converges almost everywhere, and since the $(I-1)^{th}$ partial sum of this series is precisely g_{n_I}, we find that

$$g_{n_i}(x) \to g(x) \quad a.e. \quad x.$$

Now we try to prove that $g_{n_i}(x) \to g(x)$ in $L^p(R)$; we observe that $|g - s_I(g)(x)|^2 \leq (2g)^2$ for all I and apply the dominated convergence theorem to get $\|g_{n_i} - g\| \to 0$ as i tend to infinity.

Since $\{g_n\}_{n=1}^{\infty}$ is a Cauchy sequence, given ε, there exists N such that for all $n, m > N$ we have $\|g_n - g_m\| < \frac{\varepsilon}{2}$. If n_i is chosen so that $n_i > N$, and $\|g_{n_i} - g\| < \frac{\varepsilon}{2}$, then the triangle inequality implies

$$\|g_n - g\| \leq \|g_n - g_{n_i}\| + \|g_{n_i} - g\| < \varepsilon \text{ for all } n_i, n > N.$$

Bounded functions or L^∞ functions: A piecewise continuous function $f(x)$ defined on an interval I is bounded (or L^∞) on I if there is a number $M > 0$ such that

$$|f(x)| \leq M. \quad \text{for all} \quad x \in I$$

The L^∞ norm of a function $f(x)$ is defined by

$$\|f\|_\infty = \sup\{|f(x)| : x \in I\}.$$

Example C.5:

1. The function $f(x) = \sin x$ and $f(x) = \cos x$ are L^∞ on R. Also the complex valued function $f(x) = e^{ix}$ is L^∞ on R. In fact,

$$\|\sin x\|_\infty = \|\cos x\|_\infty = \|e^{ix}\|_\infty = 1$$

2. Any polynomial function of degree greater than zero $p(x)$ is not in L^∞ on R.
3. The function $f(x) = \frac{1}{x}$ is continuous and has finite value at each point of the interval $(0,1]$ but is not bounded on $(0,1]$.

Integrable function (L^1): A piecewise continuous function $f(x)$ defined on an interval I as integrable (or of class L^1) on I if

$$\int_I |f(x)| dx < \infty \text{ i.e., finite.}$$

The L^1 norm of a function $f(x)$ is defined by

$$\|f\|_1 = \int_I |f(x)| dx.$$

In general, a function $f(x) \in L^p$ if

$$\int_I |f(x)|^p dx < \infty, \; 1 \le p < \infty$$

Question: Is L^p on interval I a Hilbert space?

Ans: Suppose $1 \le p \ne 2 < \infty$ then for any interval I,

$$\langle f, f \rangle = \int_I f(x) \overline{f(x)} . dx \ne \left(\int_I |f(x)|^p \, dx \right)^{1/p}$$

thus for $p \ne 2$, L^p is not a Hilbert space.

The sequence space $l^p (1 \le p < \infty)$: It contains sequences $z = \{z_n\}_{n \ge 1}$ satisfying

$$\sum_{n=1}^{\infty} |z_n|^p < \infty$$

The l^p norm (or simply p-norm) and supremum norm (or simply sup norm) are given by

$$\|z\|_p = \left\{ \sum_{k=1}^{\infty} |z_k|^p \right\}^{1/p} \quad \text{if } 1 \le p < \infty \text{ and}$$

The Hilbert space $H^1(\Omega)$ inner product and norm includes the functions and their first derivatives

$$(u,v)_1 = \int_\Omega \left[uv + \sum_{k=1}^{n} u_{,k} v_{,k} \right] d\Omega$$

where $(\;)_{,k} = \partial(\;)/\partial x_k$, and

$$\|u\|_{H^1} = \|u\|_1 = (u,u)_1^{\frac{1}{2}} = \left[\int_\Omega \left(u^2 + \sum_{k=1}^{n} u^2_{,k} \right) d\Omega \right]^{\frac{1}{2}}.$$

Likewise, we can extend $H^s(\Omega)$ to include the S-th order derivates.

Appendix D: Daubechies Filter and Connection Coefficients

D.1 FILTER COEFFICIENTS

Consider the refinement relations for scaling function

$$\phi(x) = \sum_{k=-\infty}^{+\infty} a_k \, \phi(2x-k) \qquad \text{(D.1)}$$

Here we have to find the filter coefficients a_k to get the following desired properties within a scaling function:

1. The area under the scaling function is normalized to unity, i.e.,

$$\int_{-\infty}^{+\infty} \phi(x)\,dx = 1 \qquad \text{(D.2)}$$

Substituting equation (D.1), we get

$$\sum_{k=-\infty}^{+\infty} a_k \int_{-\infty}^{+\infty} \phi(2x-k)\,dx = 1$$

Putting $y = 2x - k$ and using equation (D.2), we get

$$\sum_{k=-\infty}^{+\infty} a_k = 2 \qquad \text{(D.3)}$$

2. The orthogonal scaling function satisfies the relation

$$\int_{-\infty}^{+\infty} \phi(x)\phi(x+l)\,dx = \delta_{0,l} \quad l \in Z \tag{D.4}$$

or $$\int_{-\infty}^{+\infty} \sum_{k=-\infty}^{+\infty} a_k\,\phi(2x-k) \sum_{m=-\infty}^{+\infty} a_m\,\phi(2x+2l-m)\,dx = \delta_{0,l}$$

or $$\sum_{k=-\infty}^{+\infty} a_k \sum_{m=-\infty}^{+\infty} a_m \int_{-\infty}^{+\infty} \phi(2x-k)\phi(2x+2l-m)\,dx = \delta_{0,l}$$

Putting $y = 2x - k$, we get

$$\sum_{k=-\infty}^{+\infty} a_k \sum_{m=-\infty}^{+\infty} a_m \int_{-\infty}^{+\infty} \phi(y)\phi(y+2l-m+k)\,dy = 2\delta_{0,l}$$

The integral is non-zero when $2l - m + k = 0$ or $m = 2l + k$, therefore

$$\sum_{k=-\infty}^{+\infty} a_k\,a_{2l+k} = 2\delta_{0,l} \tag{D.5}$$

3. A wavelet is orthogonal to scaling function if it satisfies the following relation:

$$\psi(x) = \sum_{k=-\infty}^{+\infty} (-1)^k\,a_{N-1-k}\,\phi(2x-k) \tag{D.6}$$

It can be verified that $\langle \phi(x), \psi(x) \rangle = 0$.

If the scaling function can exactly represent polynomials then

$$\alpha_0 + \alpha_1 x + \alpha_2 x^2 + \ldots\ldots + \alpha_{p-1} x^{p-1} = \sum_{k=-\infty}^{+\infty} c_k\,\phi(x-k) \tag{D.7}$$

Since $\langle \phi(x-k), \psi(x) \rangle = 0$,

$$\alpha_0 \int_{-\infty}^{+\infty} \psi(x)\,dx + \alpha_1 \int_{-\infty}^{+\infty} \psi(x)x\,dx + \alpha_2 \int_{-\infty}^{+\infty} \psi(x)x^2\,dx + \ldots\ldots + \alpha_{p-1} \int_{-\infty}^{+\infty} \psi(x)x^{p-1}\,dx = 0$$

Each term of the preceding equation must be zero for exact representation of polynomials of order up to p but not greater than p. The first p moments of the wavelet must be zero, therefore

$$\int_{-\infty}^{+\infty} \psi(x)x^l dx = 0, \quad l = 0,1,2,...,p-1 \tag{D.8}$$

Substituting equation (D.6), we get

$$\sum_{k=-\infty}^{+\infty} (-1)^k a_{N-1-k} \int_{-\infty}^{+\infty} x^l \phi(2x-k)dx = 0$$

Let $2x = t$ then

$$\sum_{k=-\infty}^{+\infty} (-1)^k a_{N-1-k} \int_{-\infty}^{+\infty} t^l \phi(t-k)dt = 0 \tag{D.9}$$

We have to put the moment equation to simplify equation D.9. The moment is expressed in equation D.13 of this appendix as

$$\int_{-\infty}^{+\infty} t^l \phi(t-k)dt = c_k^l = \sum_{i=0}^{l} \binom{l}{i} k^{l-i} c_0^i$$

For simplicity, we use Daubechies 4 Scaling function ($N = 4$), i.e., $l = 0,1$.

Case 1: For $l = 0$

$$c_k^0 = \sum_{i=0}^{0} \binom{0}{i} k^{0-i} c_0^i = 1 \tag{D.10a}$$

Case 2: For $l = 1$

$$c_k^1 = \sum_{i=0}^{1} \binom{1}{i} k^{1-i} c_0^i = \binom{1}{0} k^{1-0} c_0^0 + \binom{1}{1} k^{1-1} c_0^1 = k + c_0^1$$

The coefficient c_0^1 can be calculated by using equation D.14 of this appendix.

$$c_0^{l=1} = \frac{1}{2(2^1-1)} \sum_{i=0}^{l-1} \binom{1}{i} \left[\sum_{k=0}^{N-1} a_k k^{1-i} c_0^i \right]$$

$$c_0^1 = \frac{1}{2} \left[a_0 (0)^1 c_0^0 + a_1 (1)^1 c_0^0 + a_2 (2)^1 c_0^0 + a_3 (3)^1 c_0^0 \right]$$

$$c_0^1 = \frac{1}{2}\left[a_1 + 2a_2 + 3a_3\right]$$

Hence,

$$c_k^1 = k + \frac{1}{2}\left[a_1 + 2a_2 + 3a_3\right] \tag{D.10b}$$

Substituting equation (D.10a) in equation (D.9), we get

$$\sum_{k=-\infty}^{+\infty}(-1)^k\, a_{N-1-k} = 0 \tag{D.11a}$$

Substituting equation (D.10b) in equation (D.9), we get

$$\sum_{k=-\infty}^{+\infty}(-1)^k\, a_{N-1-k}\, k + \sum_{k=-\infty}^{+\infty}(-1)^k\, a_{N-1-k}\left[\frac{1}{2}\{a_1 + 2a_2 + 3a_3\}\right] = 0$$

Since $a_1 + 2a_2 + 3a_3$ is a constant, using equation (D.11a), we get

$$\sum_{k=-\infty}^{+\infty}(-1)^k\, a_{N-1-k}\, k = 0 \tag{D.11b}$$

Equations D.11a and D.11b can be generalized, and it can be proved that [1]

$$\sum_{k=-\infty}^{+\infty}(-1)^k\, a_k\, k^l = 0, \ l = 0,1,2,\ldots,p-1 \tag{D.12}$$

It will yield $N/2$ equations.

The filter coefficients a_k for $k = 0,\ldots,N-1$ are uniquely defined by equations (D.3), (D.5) and (D.12).

D.1.1 Mathematica Program for Construction of Filter Coefficients for D4

$$N = 4,\ l = 0,1$$

From equation (D.3),

$$a_0 + a_1 + a_2 + a_3 = 2$$

From equation (D.5), for $l = 0,1$

$$a_0^2 + a_1^2 + a_2^2 + a_3^2 = 2$$

$$a_0\, a_2 + a_1\, a_3 = 0$$

From equation (D.12), for $l=0,1$

$$a_0 - a_1 + a_2 - a_3 = 0$$

$$-a_1 + 2a_2 - 3a_3 = 0$$

Now we have to solve this system of 5 nonlinear equations for 4 constants. We can get values of all the filter constants upon solving any 4 of these equations. Here is the Mathematica code

$$\text{NSolve} \ \ [\{a0+a1+a2+a3-2, a0\wedge2+a1\wedge2+a2\wedge2+a3\wedge2-2, a0*a2$$

$$+a1*a3, -a1+2*a2-3*a3\}, \{a0, a1, a2, a3\}, \text{Reals}]$$

and the solution is:

$$\{a0 \rightarrow 0.6830127018921976, a1 \rightarrow 1.1830127018921903, a2 \rightarrow 0.31698729810780524,$$

$$a3 \rightarrow -0.18301270189219332\}$$

D.1.2   Mathematica Program for Construction of Filter Coefficients for D6

$$N=6, \ l=0,1,2$$

From equation (D.3),

$$a_0 + a_1 + a_2 + a_3 + a_4 + a_5 = 2$$

From equation (D.5), for $l=0,1,2$

$$a_0{}^2 + a_1{}^2 + a_2{}^2 + a_3{}^2 + a_4{}^2 + a_5{}^2 = 2$$

$$a_0 a_2 + a_1 a_3 + a_2 a_4 + a_3 a_5 = 0$$

$$a_0 a_4 + a_1 a_5 = 0$$

From equation (D.12), for $l=0,1,2$

$$a_0 - a_1 + a_2 - a_3 + a_4 - a_5 = 0$$

$$-a_1 + 2a_2 - 3a_3 + 4a_4 - 5a_5 = 0$$

$$-a_1 + 4a_2 - 9a_3 + 16a_4 - 25a_5 = 0$$

Now we have to solve this system of 7 nonlinear equations for 6 constants. We can get values of all the filter constants upon solving any 6 of these equations. The code for Mathematica is.

NSolve [{a0 + a1 + a2 + a3 + a4 + a5 − 2, a0^2 + a1^2 + a2^2 + a3^2 + a4^2

  + a5^2 − 2, a0 * a2 + a1 * a3 + a2 * a4 + a3 * a5, a0 − a1 + a2 − a3 + a4 − a5, −a1

  + 2 * a2 − 3 * a3 + 4 * a4 − 5 * a5, −a1 + 4 * a2 − 9 * a3

  + 16 * a4 − 25 * a5}, {a0, a1, a2, a3, a4, a5}, Reals]

We get the solution

  {a0 → 0.4704672077841641, a1 → 1.1411169158314438, a2 → 0.6503650005262318,

  a3 → −0.19093441556832763, a4 → −0.12083220831039561,

  a5 → 0.04981749973688408}

D.2 MOMENT OF THE SCALING FUNCTION

The moment of the scaling can be expressed as:

$$c_k^l = \int_{-\infty}^{\infty} x^l \phi(x-k)\,dx$$

putting $x - k = y$

$$c_k^l = \int_{-\infty}^{\infty} (y+k)^l \phi(y)\,dy$$

$$= \int_{-\infty}^{\infty} \sum_{i=0}^{l} \binom{l}{i} k^{l-i} y^i \phi(y)\,dy$$

$$= \sum_{i=0}^{l} \binom{l}{i} k^{l-i} \int_{-\infty}^{\infty} y^i \phi(y)\,dy$$

$$= \sum_{i=0}^{l} \binom{l}{i} k^{l-i} c_0^i \tag{D.13}$$

Consider again the moment equation

$$c_k^l = \int_{-\infty}^{\infty} x^l \phi(x-k)\,dx$$

Substituting the refinement relations, we get

$$c_k^l = \int_{-\infty}^{\infty} x^l \sum_j a_j \phi(2x-2k-j)\,dx$$

putting $y = 2x$

$$c_k^l = \sum_j a_j \int \left(\frac{y}{2}\right)^l \phi(y-2k-j)\frac{dy}{2} = \frac{1}{2^{l+1}} \sum_j a_j c_{j+2k}^l$$

Let $i = j+2k$ then

$$c_k^l = \frac{1}{2^{l+1}} \sum_i a_{i-2k} c_i^l$$

Some manipulations show

$$c_0^l = \frac{1}{2^{l+1}} \sum_i a_i c_i^l = \frac{1}{2^{l+1}} \sum_k a_k c_k^l$$

Substituting equation D.13, we get

$$c_0^l = \frac{1}{2^{l+1}} \sum_{i=0}^{l} \binom{l}{i} \sum_k a_k k^{l-i} c_0^i$$

$$c_0^l = \frac{1}{2^{l+1}} \left[\sum_{i=0}^{l-1} \binom{l}{i} \sum_k a_k k^{l-i} c_0^i + \binom{l}{l} \sum_k a_k k^{l-l} c_0^l \right]$$

Since $\sum_k a_k = 2$, therefore

$$c_0^l = \frac{1}{2^{l+1}} \left[\sum_{i=0}^{l-1} \binom{l}{i} \sum_k a_k k^{l-i} c_0^i + 2 c_0^l \right]$$

It can be simplified as

$$c_0^l = \frac{1}{2(2^l - 1)} \sum_{i=0}^{l-1} \binom{l}{i} \sum_k a_k k^{l-i} c_0^i \tag{D.14}$$

with $c_0^0 = 1$

D.3 CONNECTION COEFFICIENTS

We get the following types of connection coefficients when we use Galerkin method for the solution of differential equations

$$\Omega^r[n] = \int_{-\infty}^{+\infty} \phi^r(x)\phi(x-n)\,dx; \qquad r > 0 \tag{D.15}$$

Here the superscript of scaling represents the order of the derivative with respect to x. The refinement relations can be expressed as

$$\phi^r(x) = 2^r \sum_{k=0}^{N-1} a_k \phi^r(2x-k)$$

$$\phi(x-n) = \sum_{l=0}^{N-1} a_l \phi(2x-2n-l)$$

Substituting these refinement relations in equation D.15, we get [2]

$$\Omega^r[n] = 2^{r-1} \sum_{k=0}^{N-1}\sum_{l=0}^{N-1} a_l a_k \Omega[2n+l-k] \tag{D.16}$$

This can also be written in the matrix form $2^{1-r}\{\Omega\} = [A]\{\Omega\}$. To find a unique connection coefficient, the equation is not sufficient; therefore, polynomials approximating property of Daubechies scaling functions are also used.

The connection coefficients have the important property

$$\Omega^r[n] = (-1)^r \Omega^r[-n] \tag{D.17}$$

The Daubechies scaling functions have the polynomial approximation properties of [3,4]

$$\sum_{k=-\infty}^{\infty} k^r \phi(x-k) = P_r(x)$$

where $P_r(x)$ is a polynomial of degree r for $0 \le r \le \left(\dfrac{N}{2}\right) - 1$.

Differentiating it r times, we get

$$\sum_{k=-\infty}^{\infty} k^r \phi^r(x-k) = r!$$

Multiplying both sides by $\phi(x)$ and integrating it, we get

$$\sum_{k=-\infty}^{\infty} k^r \int_{-\infty}^{\infty} \phi(x)\phi^r(x-k)\,dx = r! \int_{-\infty}^{\infty} \phi(x)\,dx$$

Putting $y = x - k$

$$\sum_{k=-\infty}^{\infty} k^r \int_{-\infty}^{\infty} \phi(y+k)\phi^r(y)dy = r!$$

We obtain the normalization conditions for connection coefficients

$$\sum_{k=-\infty}^{\infty} k^r \Omega^r[-k] = r! \tag{D.18}$$

```
% Program to calculate connection coefficients
clc
close all
clear all
% for order m
N=6;
m=input('enter value of m = ')
a(1)=0.47046721;a(2)=1.14111692;a(3)=0.650365;
a(4)=-0.1903442;a(5)=-0.12083221;a(6)=0.0498175;

A=zeros(N,N);
for n=1:N
for k-1:N
for L=1:N
jj=2*(n-1)+L-k;
if rem(m,2)~=0
if jj<6&&jj>=0;
                A(n,jj+1)=A(n,jj+1)+a(k)*a(L);

end
if jj>-6&&jj<0;
jj=abs(jj);
                A(n,jj+1)=A(n,jj+1)-a(k)*a(L);
end
else
if jj<6
jj=abs(jj);
                A(n,jj+1)=A(n,jj+1)+a(k)*a(L);
end
end
end
end
end
```

```
[V,D]=eig(A)
Z=V(:,2)

NM=0;
fori=2:N-1
      NM=NM+((i-1)^m)*Z(i);
end

CC=((-1)^m)*Z*factorial(m)/(2*NM)
```

REFERENCES

1. J.R. Williams and K. Amaratunga (2016), *Introduction to Wavelets in Engineering*, https://hal.archives-ouvertes.fr/hal-01311772. Accessed on April 22, 2019.
2. K. Amaratunga and J.R. Williams (1997), Wavelet-Galerkin solution of boundary value problems, *Archives of Computational Methods in Engineering*, Vol. 4, No. 3, pp. 243–285.
3. G. Beylkin and J. M. Keiser (1997), *An Adaptive Pseudo-Wavelet Approach for Solving Nonlinear Partial Differential Equations, Wavelet Analysis and Applications*, V. 6, Academic Press, San Diego, CA.
4. S. Goedecker (1998), *Wavelets and Their Application for the Solution of Partial Differential Equations in Physics*, Université de Lausanne Presses polytechniques et universitaires romandes, Lausanne, Switzerland.

Appendix E:
Fourier Transform

FRENCH MATHEMATICIAN JEAN BAPTISTE Joseph Fourier (1768–1830) first introduced the concept of infinite series in terms of sines and cosines to represent any function under certain general conditions. This series, named the Fourier series, is perhaps the single most important mathematical tool used in signal processing. It represents the fundamental procedure by which complex signals may be decomposed into simpler ones and, conversely, by which complicated signals may be created out of simpler building blocks. Mathematically, Fourier analysis has spawned some of the most fundamental developments in the understanding of infinite series and function approximation. Equally important, it is the tool with which many of the everyday phenomena—the perceived differences in sound between violins and drums, sonic booms, and the mixing of colors—can be better understood.

The Fourier series of a function $f(t)$ on the periodic interval $(-T, T)$ can be expressed as

$$f(t) = \sum_{n=-\infty}^{\infty} c_n e^{in\pi t/T}$$

where:

$$c_n = \frac{1}{2T} \int_{-T}^{T} f(\tau) e^{-in\pi\tau/T} d\tau$$

Substituting $\omega_n = n\pi/T$ and the coefficient into the series, we get

$$f(t) = \frac{1}{2\pi} \sum_{n=-\infty}^{\infty} \left[\int_{-T}^{T} f(\tau) e^{-i\omega_n\tau} d\tau \right] e^{i\omega_n t} \frac{\pi}{T}$$

As the period $T \to \infty$, the function becomes aperiodic, ω_n becomes a continuous variable ω. The sum becomes an integral and $d\omega = \pi/T$. Therefore,

$$f(t) = \frac{1}{2\pi} \int_{-\infty}^{\infty} \left[\int_{-\infty}^{\infty} f(\tau) e^{-i\omega\tau} d\tau \right] e^{i\omega t} d\omega$$

It can also be stated as

$$f(t) = \frac{1}{2\pi} \int_{-\infty}^{\infty} F(\omega) e^{i\omega t} d\omega$$

where

$$F(\omega) = \int_{-\infty}^{\infty} f(t) e^{-i\omega t} dt$$

$F(\omega)$ is known as the Fourier transform of $f(t)$.

Theorem: Let $\{\varphi_n\}$ be a countable infinite orthonormal set in a Hilbert space H. Then the following statement holds:

The infinite series $\sum_{n=1}^{\infty} \alpha_n \varphi_n$, represent a function that belongs to Hilbert space where α_n are scalars, and $\sum_{n=1}^{\infty} |\alpha_n|^2$ converges.

In case of Fourier series $\varphi_n = e^{i\omega_n t}$.

The following three notions are commonly used to test the convergence or the closeness of the representation with the exact function.

1. Convergence is called pointwise convergence if a series $\sum_{n=1}^{N} c_n \varphi_n$ converges to $f(x)$ for each $x \in (a,b)$ such that

$$\left| f(x) - \sum_{n=1}^{N} c_n \varphi_n \right| \to 0 \quad \text{as} \quad N \to \infty.$$

2. Convergence is called uniform convergence or the series converges uniformly to $f(x)$ for each $x \in [a,b]$ if

$$\max_{a \leq x \leq b} \left| f(x) - \sum_{n=1}^{N} c_n \varphi_n \right| \to 0 \qquad \text{as} \quad N \to \infty$$

3. Convergence is called L2 convergence or the series converges in the mean-square sense to $f(x)$ in (a, b) if

$$\int_{a}^{b} \left| f(x) - \sum_{n=1}^{N} c_n \varphi_n \right|^2 dx \to 0 \quad \text{as } N \to \infty$$

Note that boundary points are included in the case of uniform convergence but not in the other two cases. The type of convergence of a series depends on the function $f(x)$ and its continuity.

(Parseval Equality) For any pair of vectors $u, v \in H$, one has

$$\langle u, v \rangle = \sum_i \langle u, \varphi_i \rangle \langle v, \varphi_i \rangle$$

For any $u \in H$, one has

$$\|u\|^2 = \sum_{i=1} |u, \varphi_i|^2$$

Any linear subspace S of H that contains $\{\varphi_i\}$ is dense in H.

Theorem (Plancherel theorem): Parseval's equality expresses the energy of a system in time and frequency domain [1]. Let $f(t) \in L^2(R)$ then,

$$\int_{-\infty}^{\infty} |f(t)|^2 \, dt = \frac{1}{2\pi} \int_{-\infty}^{\infty} |F(\omega)|^2 \, d\omega$$

Also,

$$\int_{-\infty}^{\infty} f(t) \overline{g(t)} \, dt = \frac{1}{2\pi} \int_{-\infty}^{\infty} F(\omega) \overline{G(\omega)} d\omega$$

For the normalized wave function $f(t)$,

$$\int_{-\infty}^{\infty} |f(t)|^2 \, dt = 1$$

E.1 HEISENBERG UNCERTAINTY PRINCIPLE

Our mind accepts the information in the domain of time and space, e.g., continuous formation of images in eyes while reading. It also works in the frequency domain, e.g., hearing where we recognize the frequency of sound vibrations. While Fourier provided a method to transform time-space function into function of frequency domain, Heisenberg established a relation regarding simultaneously perceiving both time and frequency information of any signal or vibration. This well known uncertainty principle uses Fourier transform, which is an extension of a Fourier series for aperiodic function.

Theorem: Let $\bar{T}^2 = \int_{-\infty}^{\infty} |tf(t)|^2\, dt$ and $\bar{\Omega}^2 = \frac{1}{2\pi}\int_{-\infty}^{\infty} |\omega F(\omega)|^2\, d\omega$ be the expected values for a test function $f(t)$. The uncertainty principle asserts that

$$\bar{T}.\bar{\Omega} \geq \frac{1}{2}$$

Proof: By Schwarz's inequality, we have

$$\left| \int_{-\infty}^{\infty} tf(t)f'(t)dt \right| \leq \left[\int_{-\infty}^{\infty} |tf(t)|^2\, dt \right]^{\frac{1}{2}} . \left[\int_{-\infty}^{\infty} |f'(t)|^2\, dt \right]^{\frac{1}{2}}$$

Consider the right side of the equation. By definition $\frac{df}{dt} = i\omega F(\omega)$ and by Parseval's equality

$$\bar{T}.\left[\int_{-\infty}^{\infty} |i\omega F(\omega)|^2 \frac{d\omega}{2\pi} \right]^{\frac{1}{2}} = \bar{T}.\bar{\Omega}$$

Integration by parts of the left side of the equation

$$\int_{-\infty}^{\infty} t\left(\frac{1}{2}\frac{df^2}{dt} \right) dt = \frac{1}{2}t[f(t)]^2 \Big|_{-\infty}^{\infty} - \int_{-\infty}^{\infty} \frac{1}{2}[f(t)]^2\, dt = 0 - \frac{1}{2}$$

Therefore,

$$\frac{1}{2} \leq \bar{T}.\bar{\Omega}$$

This relation is useful for wavelets in time-frequency analysis which shows as the accuracy of time information improves, the frequency of information deteriorates. If we write expected frequency in Hz, then the above equation can be expressed as:

$$\frac{1}{4\pi} \leq \bar{T}.\bar{f}$$

E.2 CONVOLUTIONS USING DISCRETE FOURIER TRANSFORM

In the case of continuous function, the Fourier series can reproduce the periodic function $u(x)$ over an interval by using sines and cosines (or exponentials) as the basis functions. A combination of exponentials e^{ikx} rather than $\sin(kx)$ and $\cos(kx)$ is preferred for the discrete Fourier series. In the case of n discrete values, n linear algebraic equations are used to find the Fourier coefficients $c_0,\ldots\ldots c_{n-1}$. The n discrete values of u_i are considered at equal spaces and equated with the finite series $c_0 + c_1 e^{ix} + c_2 e^{2ix} + \ldots$($n$ terms) between 0 and 2π with the spacing $2\pi/n$. The jth discrete value will be at $x = (j-1) \times (\frac{2\pi}{n})$. If we use $w = e^{2\pi i/n}$, then the discrete value can be expressed as

$$at\ x = (j-1)*2\pi/n \qquad c_0 + c_1 w^{(j-1)} + c_2 w^{2(j-1)} + \cdots\cdots + c_{n-1} w^{(n-1)(j-1)} = u_{j-1}$$

Therefore,

First value $(j=1$ or at $x=0)$: $c_0 + c_1 + c_2 + \cdots\cdots + c_{n-1} = u_0$

Second value $(j=2$ or at $x=2\pi/n)$: $c_0 + c_1 w + c_2 w^2 + \cdots\cdots + c_{n-1} w^{n-1} = u_1$

Third value $(j=3$ or at $x=4\pi/n)$: $c_0 + c_1 w^2 + c_2 w^4 + \cdots\cdots + c_{n-1} w^{2(n-1)} = u_2$

In general $w = e^{2\pi i/n}$ lies on the unit circle $|w|=1$ in the complex plane, at an angle $2\pi/n$ from horizontal and $w^n = 1$. All the entries of the first row $(j=1)$ and the first column $(n=1)$ are $w^0 = 1$, and the other entries are $w^{(j-1)(n-1)}$. The n equations can be expressed in the matrix form as

$$
\begin{bmatrix}
1 & 1 & 1 & . & 1 \\
1 & w & w^2 & . & w^{n-1} \\
1 & w^2 & w^4 & . & w^{2(n-1)} \\
. & . & . & . & . \\
1 & w^{n-1} & w^{2(n-1)} & . & w^{(n-1)^2}
\end{bmatrix}
\begin{bmatrix}
c_0 \\
c_1 \\
c_2 \\
. \\
c_{n-1}
\end{bmatrix}
=
\begin{bmatrix}
u_0 \\
u_1 \\
u_2 \\
. \\
u_{n-1}
\end{bmatrix}
$$

The equation $T_f c = u$ can be inverted to find c_i. The inverse of matrix T_f can be expressed in terms of its complex conjugate $\bar{T}_f$ as [2]

$$
T_f^{-1} = \frac{\bar{T}_f}{n} = \frac{1}{n}
\begin{bmatrix}
1 & 1 & 1 & . & 1 \\
1 & w^{-1} & w^{-2} & . & w^{-(n-1)} \\
1 & w^{-2} & w^{-4} & . & w^{-2(n-1)} \\
. & . & . & . & . \\
1 & w^{-(n-1)} & w^{-2(n-1)} & . & w^{-(n-1)^2}
\end{bmatrix}.
$$

For the input sequence $f_i = 1,3,5,8$, the output sequence c_0, c_1, c_2, c_3 can be expressed as (for $n=4$, w is i)

$$
c_0 + c_1 + c_2 + c_3 = 1
$$
$$
c_0 + ic_1 - c_2 - ic_3 = 3
$$
$$
c_0 - c_1 + c_2 - c_3 = 5
$$
$$
c_0 - ic_1 - c_2 + ic_3 = 8
$$

The coefficient matrix T_f can be written as

$$
T_f =
\begin{bmatrix}
1 & 1 & 1 & 1 \\
1 & i & i^2 & i^3 \\
1 & i^2 & i^4 & i^6 \\
1 & i^3 & i^6 & i^9
\end{bmatrix}
$$

The complex conjugate of this matrix can be obtained by replacing i by $-i$.

$$\bar{T}_f = \begin{bmatrix} 1 & 1 & 1 & 1 \\ 1 & (-i) & (-i)^2 & (-i)^3 \\ 1 & (-i)^2 & (-i)^4 & (-i)^6 \\ 1 & (-i)^3 & (-i)^6 & (-i)^9 \end{bmatrix}$$

And $T_f^{-1} = \dfrac{1}{4}\bar{T}_f$

The coefficients c_i can be easily calculated using conjugate matrix as

$$c_0 = (1+3+5+8)/4$$

$$c_1 = (1-3i-5+8i)/4$$

$$c_2 = (1-3+5-8)/4$$

$$c_3 = (1+3i-5-8i)/4$$

Let us consider two vectors $f = (f_0, f_1, \ldots, f_{n-1})$ and $g = (g_0, g_1, \ldots, g_{n-1})$ of length n. Every i^{th} term ($i = 0,1,\ldots,n-1$) of this convolution contains all products $f_j g_k$ with $j+k=i$ or $j+k=i+n$. The convolution product $(f*g)$ of the two vectors will be

$$f*g = (f_0 g_0 + f_1 g_{n-1} + f_2 g_{n-2} + \ldots + f_{n-1} g_1, \ldots, \ldots, f_0 g_{n-1} + f_1 g_{n-2} + \ldots + f_{n-1} g_0)$$

Convolution of two vectors is more complicated than just multiplication. The first term of the convolution is the sum of the product of all f_j and g_k such that $j+k=0$ or $j+k=n$. Similarly, for the second term, the sum is $j+k=1$ or $n+1$, and so on.

Let us take an example: convolution of $(2,3,4)$ and $(3,2,1)$ is

$$(2\cdot3+3\cdot1+4\cdot2, 2\cdot2+3\cdot3+4\cdot1, 2\cdot1+3\cdot2+4\cdot3) = (17,17,20).$$

Or we can also find it by multiplication

$$(2+3w+4w^2)\cdot(3+2w+w^2) = 6+13w+20w^2+11w^3+4w^4$$

Here, $(f*g)$ should have 3 components instead of 5 and as $n=3$, so $w^3=1$, $w^4=w$. Then, the product becomes $(17+17w+20w^2)$, which is the same as before. It can also be calculated using fast Fourier transform (*FFT*).

Let c and d be the Fourier transform of f and g, respectively, i.e., $c = \frac{1}{n}T_f^{-1}f$ and $d = \frac{1}{n}T_f^{-1}g$. The convolution of the vectors $(f*g)$ can be computed by multiplying the c and d components by component $(c_i.d_i)$ and then transforming back: $f*g = n.T_f(cd)$. The algorithm developed for discrete Fourier transform by Cooley and Tukey known as FFT is a very efficient technique for computing transform. This process of computing convolution looks long because three times FFT are used, but it is faster than the process described earlier. It saves a significant amount of time for a large data set. It is to be noted that the functions are assumed to be periodic.

REFERENCES

1. W. Miller (2006), *Introduction to the Mathematics of Wavelets*, University of Minnesota, Amsterdam, the Netherlands.
2. G. Strang (1986), *Introduction to Applied Mathematics*, Wellesley-Cambridge Press, Wellesley, MA.

Index

Note: Page numbers in italic refer to figures and those in bold to tables.